Seal Doctor

Before Ken Jones moved to Cornwall he had rarely seen the sea and hardly knew seals existed. Born in Tonypandy in the Rhondda valley at the time of the Depression, he was forced to leave school early and go down the mines to help support the family. He left Wales at the age of twenty to become a trainee nurse at Burntwood Hospital in Staffordshire, but went back to mining again (for the better pay) on getting married. A bad accident and the policy of closing down old pits decided him to invest his savings in a beach café at St Agnes, Cornwall. It was there, at Trevaunance Cove, that he met his first 'orphan of the sea', a two-day-old seal pup who refused to be put back to sea. Ken and his wife Mary believe that fate led them to a life that neither could have foreseen in their wildest dreams, scrambling down cliffs on stormy winter nights to rescue seals battered on the rocks, learning how to nurture pups and care for those that had been injured.

1969 proved a disastrous winter for seals and sea-birds around the Cornish coast. Because of the Torrey Canyon disaster and the resultant oil pollution, press and television covered the story in a big way and the Jones's café and bungalow became a centre for the media as well as a hospital for sea-birds. Despite tremendous efforts and care very few seals were saved. It was this experience that determined Ken to build a proper seal sanctuary. After six years of trial and tribulation, plus opposition from some quarters, he at last realized his dream when at Whitsun, 1975, he welcomed the first visitors to his new Seal Sanctuary at Gweek on the Helford River.

KEN JONES

SEAL DOCTOR

FONTANA/COLLINS

A portion of this book was first published as
Orphans of the Sea by Harvill Press 1970
First published in Fontana Paperbacks 1978
Fourth impression, with Afterword, March 1986
Fifth impression October 1988

Printed and bound in Great Britain by
William Collins Sons & Co. Ltd, Glasgow

FOR MARY AND LINDA

Contents

Publisher's Note

Ken Jones' account of his early work in rescuing and caring for the seals that were washed ashore near his home in Cornwall was told in his book *Orphans of the Sea*, first published by the Harvill Press in 1970 and subsequently reprinted many times. This book took the story of his first Seal Sanctuary at St Agnes up to March 1970, when oil pollution and disease were wreaking havoc amongst – and exciting popular concern for – the wild life to which Ken Jones had devoted his life. Since that time, the Sanctuary has grown considerably in size, scope and reputation, and has moved from its original site to an entirely new location at Gweek, near Helston in Cornwall. It is now not only a unique and highly valuable institution, but also one of the best known and most popular wild life sanctuaries in the entire country, attracting many thousands of visitors each year. It was high time, therefore, that thought be given to producing a new edition of *Orphans of the Sea* which would both retain the qualities of the original book and bring the story completely up to date. *Seal Doctor*, therefore, includes both the complete text of the first book and ten completely new chapters on subsequent events for the benefit not only of the many people who read *Orphans of the Sea* or visited the St Agnes Sanctuary themselves, and would like to know what subsequently became of it, but also of readers coming afresh to this remarkable foundation on the Cornish coast and the remarkable man whose energy and concern made it possible.

Acknowledgements

I wish to thank all those who have supplied photographs which have been used to illustrate this book. In particular: Bob Salmon, Plymouth, Devon; Richards Brothers, Penzance; J. H. Bottrell, Penzance; Ken Young, St Agnes; Roy Hughes, St Austell; *The Sunday Mirror*.

My grateful thanks also to: Mr Griffiths of the R.S.P.C.A., Penzance, for his great efforts at the time of the Seal Disaster and for his continued interest in our work; Mr Gardner, Chief Inspector, R.S.P.C.A., Truro Branch, for his help and assistance; Veterinary Surgeons: Mr Greene of Newquay, whose advice has been invaluable; Messrs Hill and Littleton of Truro for their quick attention and patience in treatment; The Seal Research Unit, Lowestoft, Suffolk, in particular Mr Nigel Bonner and his assistant who have shown great interest in the sanctuary and who are supplying vital information; Miss Hilda Bamber of BBC Television, Plymouth, who brought the work of the Seal Sanctuary to the attention of a wide public; The *Birmingham Post* and their readers, many of whom sent donations and encouraging letters: The staff and children from schools in Cornwall, Cambridge, the Midlands and Holland, who by their activities raised funds to help seals in distress; British Cod Liver Oils (Hull and Grimsby) Ltd. for donations of their 'Super Solvitax Cod Liver Oil' used in the treatment of seals. And last but not least, our good friend and neighbour, Miss Beams Burdett, who has shown her devotion to seals on many occasions, helping with their feeding and in the Sanctuary generally.

My special thanks to the late Sir William Collins for his interest and encouragement.

It is impossible to acknowledge everyone individually who has given help or shown interest in our work for seals, but our thanks go out to them and we promise to continue to do our best to give life to the orphans of the sea.

CHAPTER ONE

A duel by television

One dark winter's day in 1972 I left my home in St Agnes to drive twenty-four miles across the Cornish countryside to the small village of Gweek by the side of the beautiful Helford River. Television cameras would be out to record my confrontation with a local group calling themselves the Helford River Protection Society who had succeeded temporarily in bringing to a halt the building of my new Seal Sanctuary. I was no stranger to the TV screen but I knew that this appearance was to be the most fateful one of my life. Should the verdict go against me I would be bankrupt and it would be the end of thirteen years' devotion to the rescue and nursing of sick and injured seals.

The car skirted the pools I had built in front of my bungalow. Every year they had been getting more and more crowded as newly-rescued babies joined the long-time resident seals. The new babies needed constant attention and I hated leaving even for a few hours. What would happen to them if I was forced to give up my work? It didn't bear thinking about. By my side was my wife, Mary, her presence bringing calm and comfort as always. All through the years she had been the greatest support a man could have, helping in every way in the care and feeding of our growing family of seals as well as looking after our beach café and bungalow, never complaining as our savings were swallowed up to fuel my determination to build the first properly equipped Seal Sanctuary in Cornwall. I could tell that underneath the outward show of calm confidence she was displaying for my benefit she was as desperately worried and nervous as I was. As we passed the beach our glances met, our thoughts going back to just such another winter's day in 1959 when we had seen our first baby seal. This was to be an event which would change our lives.

On that day, very soon after we had arrived from the Midlands to settle in St Agnes, we had noticed a crowd gathered on the beach. When we pushed our way through to the shore we saw they were watching a white, furry creature rolling with the tide, large waves pounding it against the rocks. It was about three feet long and thirty pounds in weight. We waded into the sea and got close. Dark eyes in a beautiful face looked up at us piteously. One of the eyes was cut and beneath the water we could see the umbilical cord still hanging from the creature's tummy. A newly-born seal – the first of many which would come our way subsequently.

Just as fate seemed to have led us to our work looking after seals so it seemed to have a hand in leading us to Gweek. The last thought in my mind when I began seriously looking for a site was an inland beauty spot. At that time eight acres in the sand dunes near Perranporth seemed ideal, especially as the owner was very sympathetic towards our work. My plans were passed after the usual delays, we had all our estimates done and were within a week or so of commencing work when I was told to hold back while discussions took place between the owner and the Planning Officer regarding extensions to a Holiday Camp. The next thing I heard was that my plans had been revoked. We would have to start all over again looking for a site.

After combing Cornwall, I happened upon a small farm at Gweek, in a woodland site bordering the Helford River. There was a tidal creek so it meant that sea water was available for filling the pools and the ground sloped down to the creek so that there was no possibility of flooding in heavy rain. My excitement grew as the purchase went through without a hitch and I applied for permission to build my Sanctuary at the side of the river, hidden by trees from passing boats. As the area was a beauty spot I duly had discussions with the Local Amenity Society and the Helford River Group. They decided to raise no objections and my plans were passed after some alterations to my original lay-out. Work was due to commence on 1 October, 1972.

Finance was a problem; the proceeds from the sale of my beach café and home at St Agnes would nowhere near cover the cost, but we had been successful in qualifying for a loan from the English Tourist Board. Even so, we still had to raise the rest of the money through a bank loan. At first I had intended to continue running the new sanctuary on the same voluntary basis as I was doing at St Agnes but it was made very clear to me at the various banks I approached that unless I made a small entrance charge I could never hope to get the loan I needed. Little did I realize that it was this factor which was to give rise to one of the most wounding attacks made against me, that I would be using wildlife for commercial gain.

A wave of indignation and anger flooded through me as I thought of the opposition's press campaign leading people to believe that we were plucking seals out of their natural habitat merely to display them for money to visitors. Some of the most vociferous members of the Helford River Protection Society should have been out with me on the cliffs in a howling gale, carrying thirty pounds of seal up a one-in-three slope where there was barely room for a foothold and a sheer drop hundreds of feet below to a raging sea. I caught a warning glance from Mary and relaxed my tight grip on the steering

wheel. After all, without practical experience, it is difficult to comprehend the hazards which affect seal pups
in the first few weeks of their life. In a bad winter
twenty-foot waves are smashed against the rugged
cliffs by gale force winds, washing the babies away
from their mothers. They face the fate of starvation or
being eaten alive by seagulls. Even in comparatively
good weather many of the seal pups washed in on the
beaches are 'weaners' who have been unable to learn
to catch fish. Some have smashed jaws, gashes on the
body, damaged flippers or have lost eyes. Even those
without injury are so thin and weak that most have
pneumonia or lung congestion due to exposure and starvation. All would die without help.

As our family had grown over the years so had the
fish bills (seals consume about 12/14 lbs of fish a day
when grown), the cost of heat and light, vet's fees,
special medicines for the sick and injured and the cost
of driving miles picking up stranded babies. I had
defrayed all these expenses out of my own pocket and
the small profits from our beach café – the only outside help being little donations from time to time from
well-wishers interested in our work. To start building
the new Seal Sanctuary I had mortgaged all our worldly
goods, our future livelihood. With the planning permission in my pocket I had gone ahead and spent £40,000
of borrowed money on acquiring the land and on pre-
liminary construction. If my opponents won the day,
how would I ever be able to pay it back? Lost in these
gloomy thoughts I hardly noticed that we were near-
ing the end of our journey. My stomach tightened into
a knot and my blood pressure began to rise as I prepared for the confrontation ahead. As we entered
Gweek Mary squeezed my hand. Somehow I turned
on an encouraging smile. We drove into the small
village of very old Cornish houses split by the Helford

River, past the post office and the pub to the village centre where four roads meet, the main one bridging the river. Near the bridge we turned off into the road passing alongside the creek, skirting round the new housing estate and branching off into the site where I had started building the car park. It had been designed so that all traffic visiting the Sanctuary would come via the main road and only need to travel about 150 yards along the creek road, thus causing no inconvenience or annoyance to the people on the housing estate.

The site looked in a sorry state, for the workmen had had to leave before all the messy work had been completed. Although they had only put in a month's labour the project had been going so well that I had been confident that we would be able to meet our target of opening by Easter. The forcing of all work to a standstill by the opposition's action on 29 October came like a bolt out of the blue. Top soil which had been removed in excavating the ground for the pool sites was now lying in piles all around. I knew that photographs had been taken from the air of all this temporary mess and were being used as evidence to show the spoliation of the Helford River amenities. My heart sank.

Our opponents were in the main 'foreigners', the term the Cornish people use to designate those who come from other parts of Great Britain to retire or build holiday homes. But they were a powerful and formidable group who had waged a very clever press vendetta. Now they had called in the television cameras to witness – as they hoped – their final victory. As our car rounded the final bend it was brought to a halt by a large crowd blocking the road. They were waving banners and shouting. Slowly we got out of the car, prepared to go down fighting. A great cheer went up. We could hardly believe our ears. Or our eyes as we read the messages on the banners, 'Up the seals' and 'Save our seals'. For the first time that day we smiled, then laughed and the laughter brought tears. The people who pressed around us, shouting encouragement as we made our way to the BBC television unit, were Cornish men and women, true residents of Gweek and the surrounding countryside who had been stirred into action by the thought of 'foreigners' presuming to think and act for them in opposing a Sanctuary for the saving of wildlife. The seals were 'Cornish seals', part of their heritage.

On the opposite bank of the river we could see the opposition being televised by Westward TV. On our side the BBC cameras started rolling as the interviewer asked me the first question. The duel by television had begun.

Both television channels had filmed a number of programmes over the years about our work with seals. 'Nationwide', 'Blue Peter' and 'Magpie' had given us extensive coverage and we had appeared on radio programmes like 'Woman's Hour' and 'Down Your Way'. In fact, I think I could say without boasting that, in a small way, I had become something of a celebrity in the West Country. So, the tremendous storm of opposition that had grown up at Gweek was all the more devastating. I had been getting terribly depressed over the past months, as near to a mental and physical breakdown as I had ever been in my life. I had tried not to let any of this show and to keep the worst of my worries from Mary but I often caught her watching me closely and knew that she was beginning to be afraid that my health would crack up. Now I felt good, the Cornish people rallying to my support had given me a much needed boost. There had never been any question but that I was going to fight all the way but I was in stronger heart to do so.

The BBC interviewer asked the local people whether

they thought that a Seal Sanctuary would spoil the amenities of the area as the objectors maintained. Some of the older inhabitants (who had lived in Gweek for over fifty years) said that with a coal-yard and two boat-building yards Gweek had always been an industrial area so how could five landscaped pools, hidden by trees, be detrimental? Trade was slack in the village so they welcomed the business visitors to the Sanctuary would bring. The idea that the seals would be noisy and keep them awake at nights was laughed at, as the boat and coal yards were banging away all the time and with the Culdrose helicopter base close by the air was filled with the sound of planes and helicopters night and day; it was a fact, as I pointed out, that seals make hardly any noise in captivity. But best of all they strongly asserted that the objectors were not at all representative of the Gweek residents being, in the main, a small body of newcomers, most probably visitors to Cornwall themselves in the past who, now that they had settled in the area, wanted to keep other visitors away.

The interview progressed to my work and my reasons for building the Sanctuary. For months the opposition had had the run of the local press all to themselves – next evening, when the programme went out, the true facts would at last be getting an airing and a major stage in the battle for the seals had been won. So let us go right back to the beginning.

CHAPTER TWO

How we became involved with seals

I WAS born in 1926 at Tonypandy in the Rhondda
Valley, South Wales. My father had been incapacitated
by injuries he had received in the First World War and
so I had to leave my Secondary School early and take
a job down the mines to support the family. When my
father died, at the age of forty-two, I became the bread-
winner. As a small child I was always bringing home
strayed and injured animals and although times were so
bad, my parents were sympathetic and never dis-
couraged me.

At the age of twenty I left Wales for the Midlands
and joined the staff of Burntwood Hospital as a trainee
nurse. It was in Staffordshire that I met my wife; then,
because we were both eager to get married, I left the
hospital for a better paid job in the Cannock mines. I
now studied to become a mining engineer, and after a
few years, had progressed to the status of an Official of
the Nationa: Coal Board. In 1960, owing to the closure
of many of the old mines and the effects of a bad
accident and illness, I decided to invest my savings in a
beach café at St Agnes. As soon as I had settled in
Cornwall I joined the Civil Defence, Auxiliary Coastguard Watch and Cliff Rescue services.

My wife had very much the same childhood as I; she
too was born in 1926, her father died early and she was
left with her mother and younger brother to support.
Like me she loved animals. In the early days of our
marriage she helped me in my studies and today gives
me the greatest support in my work in the Sanctuary.
Our daughter Linda was born in 1953; she too is a

great animal lover.

St Agnes is a pretty little bay situated between Perranporth and Porthtowan, twelve miles from Newquay on the one side and twenty-five miles from St Ives on the other.

Our café was in Trevaunance Cove; to reach it you go through the quaint little village and down a valley of green grass and trees. We lived just above the sea overlooking the beach, with views across to Perranporth and Newquay, and the fine rugged coastline that is typical of Cornwall. Our bungalow when we bought it was surrounded by lawns; little did we think then that all these lawns would disappear and a 'Seal Sanctuary' develop to take their place. Coming from the Midlands the sea meant little to us, only very rarely had we seen it, and we didn't even know that seals existed.

It must have been the hand of fate that decided us on buying the small beach business, because, but for this, we would not have been on the spot when our first orphan of the sea came into our bay looking for help.

She was a creamy white furry bundle, three feet long and weighing about thirty pounds. Her umbilical cord was newly cut which meant that she was only a day or two old. Many people gathered round the pup, some dogs were also interested and it was getting frightened. We left the café and went down to the bottom of the cliff. The little seal snarled as we came near. Our first instinct was to get it back to sea in case the dogs should attack it, so we got behind it and the pup started flapping across the beach; this was what we wanted it to do. Already it had cut its eyes from being washed against the rocks. When it was about thirty yards away from the cliff, we pushed it into the sea, but back it came. We repeated this operation many times until finally we decided to take the pup up to our bungalow for safety.

Several visitors followed us and we got them to carry some sea water to fill our bath. This shows you how much knowledge we had of seals. We were sure the pup must be kept in water or it would die. When the bath was full we put the little seal into it. All went well for a few seconds, then there was a splashing and suddenly the water went down the plughole and the pup lay in the empty bath looking sorry for itself. I decided I needed help and rang the nearest zoo and several animal societies to find out what I ought to do. After many telephone calls, I was told to put the pup back to sea, near to Seal Cove, which was the seals' breeding ground.

The boat was launched and we went to Seal Cove, where we threw the pup overboard, thinking it would swim away; instead it tried to climb back into the boat. We kept pushing it away, finally it swam off, and we went back to the beach. We were sitting down, smoking and laughing about our experiences, when in came a passer-by to tell us that the seal was back on the beach.

Down we went, and there it was, the same pup looking up at us with its beautiful, big, pleading eyes, as if it were saying, 'Don't do that again, I need help.' I picked it up and carried it into an old caravan in our garden. I am sure the pup thought I was her mother; whichever way I moved, she followed me. The poor thing must have been starving, so we now had to find out how to feed her and what to feed her on. Again, I made many telephone calls but got no information, until finally we were told to give her twelve ounces of margarine to one pint of milk four times a day. We were warned that more than likely we would need to force feed by a tube into the pup's stomach. The son of a friend of ours was a doctor, and he very kindly came down to show us how to do this.

When he did it it looked easy, so for the next feed we mixed the margarine and milk, and went quite confidently to the caravan. I held the pup and tried to open its jaws to put the tube down, but there was nothing doing. At first I could not move its jaw at all, but finally, I managed to force a little opening, and my wife pushed the tube through; as she did so, down came the seal's small teeth on to the tube. Again I forced its jaws open and Mary pushed the tube down, then she poured the milk into the funnel. Nothing happened, no milk was going down, the pup's teeth were again clenching the tube, so now I had to keep its jaws open until all the milk was down. Mary then pulled the tube out, and that was our first feed done. The pup gave a few hiccups but seemed contented. Mary took the feeding apparatus to the bungalow for sterilizing before the next feed. We knew that cleanliness was as essential for a seal pup as for a human baby. I stayed with Cindy, as we called her, and gave her a little fuss; she appreciated this and sucked at my hand for comfort.

The weather was so warm and sunny that after giving her a few more feeds, we took Cindy out on to the lawn. Being inquisitive, she first nosed around, then she came to my feet. I lay down on the lawn and she crawled on to my chest, sniffing at my face and scratching me with her hand-like flipper. She had no fear of me and pawed just like a dog asking to play. I trusted Cindy and she trusted me. I tickled her tummy, which she enjoyed, opening her mouth as if laughing and asking for more. On land it is difficult for a seal to play with a ball, but Cindy was soon scratching at one and trying to bite it.

I used to talk silly talk to her as if I were playing with a baby. She loved it and would turn her head to one side, as though she understood what I was saying. After playing for a while we took her back to the caravan for a rest; after all, she was only a baby and food and rest were the main essentials for her. Occasionally I popped in to keep her company. She seemed quite happy with the quantity of milk we were giving her. But we had read that seal pups wean themselves at three to four weeks old, so when she was approaching that age, I rang up some fishermen at St Ives to order fish for her.

It was Friday, and Cindy's fish was due to be delivered on the following day. I went into the caravan and greeted her, but she showed no interest. Thinking she was tired I left her to rest. We prepared the last feed of milk, the next one would be the fish, and took it to the caravan. Still she showed no interest. As, however, she had been having regular meals missing one would not matter much.

Next morning I got up early to see how she was. I went to the caravan and found her lying quite still in the corner. Thinking she must be fast asleep, I called out to her. When I did this she usually threw up her head quickly, but this time nothing happened. I crept up to her and went to tickle her tummy, she was stiff, she was dead. I could not believe it, tears came into my eyes, I gave her artificial respiration, but I could see it was too late. What had I done wrong? Why had this happened?

I sat there for quite a while, then I thought I must break the news to Mary, and to Linda who had helped with Cindy and loved her as I did. I went back to the bungalow, to the kitchen where Mary was working and stood and looked at her for a moment, then the sad news had to be told. Mary asked the same question I was asking myself: what had we done wrong for this to happen after four weeks?

I rang the vet and told him that I wanted a post-mortem. After it had been done we were informed that the margarine in the milk had proved too much for the pup to absorb. We had been given the wrong advice, we had lost Cindy through lack of knowledge. We determined to read all the books we could find on the grey seal, then, should another pup come ashore for help, we would know what to do and there would not be another tragedy. However, very little information on the feeding of pups seemed to be available, for only in recent years have any surveys, or scientific research, been done on grey seals.

CHAPTER THREE

Sammy

OUR second seal was washed in one month later. It had lost its white coat, was about four weeks old and very thin and exhausted. If it had not been so thin, I would not have taken it from the beach, as, after all, the sea was its home.

I carried it up to the caravan and put it down to rest for a while. A few hours later, I went to see how it was. As I opened the door, it came at me like a wild bull, showing its full set of sharp teeth. I closed the door, but was still able to see it through the wire netting I had put on the top half. Then I walked back, and told Mary how affectionate our new arrival was. Nevertheless, it was necessary to get fish quickly and try to feed it, so I went to the nearest fishmonger's and bought some sprats.

Carrying the fish we both approached the caravan. I opened the door and asked Mary to hold it ajar in case I needed to make a quick retreat. Then, slowly, I walked towards what looked to me like a vicious lion. I was waving a sprat in the air and gradually getting closer to it. Suddenly the seal's neck seemed to stretch out to at least two feet, I nearly lost my fingers, and the sprat was tossed across the caravan. Another quick movement from the seal and I was back through the door.

We had to find some other way of feeding our new patient. I got a stick, made a slit in the end and put a sprat into it. The seal snapped at the sprat and threw it to one side; we tried again, this time it chewed at it, and down it went. We did this with quite a few sprats, and luckily the pup ate most of them; we decided to repeat this manoeuvre four times a day.

The following morning we braved the pup, opening the door of the caravan so that we could feed it by hand. But now we were dressed for the part: wellington boots, thick jackets and gloves. Bravely I went inside, again Mary held the door open ready for a quick escape. I took the fish, a mackerel, by its tail, this at least left the length of the fish between the seal's teeth and my hand. He took it with a snatch, but not so viciously as before. Quickly I got more fish ready, and without being chewed or tasted, they went down. But when there were none left, the pup came snarling at me. I passed Mary like a sprinter, she closed the door quickly, and we sighed with relief. At least our orphan, whom we named Sampson, had had a good meal.

Things went on very much the same way for the next few days, and gradually the pup put on weight. We thought we would give it four more days – ten days in all, after that he should be able to fend for himself.

On the Saturday afternoon we took Sampson back to the beach he had come in on, put him on the water's edge, and off he went. Only twice did his head pop up, as if to say 'Thank you', and that was the last we saw of him. At least we had fattened him up a little, given him strength and weaned him on to fish, so making it possible for him to survive at sea.

In its natural life, a pup's mother leaves it after about three weeks. Swallowing a fish for the first time often proves a difficulty. Some pups do not understand what they have to do and therefore die of starvation. But the majority usually wean themselves in time, and having developed a thick layer of fat from the mother's rich milk, they have a period of a few weeks, living on this fat, in which to learn how to feed themselves.

It was in September of the following year that Sammy

the seal was washed in. After getting a telephone call, we put a clothes basket in the back of the car, in order to make it easier to carry the pup off the beach. We were expecting a gentle furry white pup which would not cause much trouble, but when we arrived we found it had shed three parts of its coat, had some nasty cuts on its head and body, and its nostrils were thick with mucus which made its breathing very bad. As we approached it snarled and waved its flipper.

I grabbed hold of its back flippers and gently pulled it into the basket, putting a blanket over it to prevent it from getting out. We took it straight to the vet for treatment, then put the basket on the back seat and Mary sat with it, in case it should decide to get out. We did not want to end up in the ditch, so I told Mary to shout if it got restless and I'd stop the car. Once or twice it tried to leave its basket, but we got home safely.

For this pup we used one of the chalets on the beach, there was more room there, and we thought the sound of the sea might make it feel at home. Owing to its condition and also its age (two to three weeks) we thought we should give it milk for a week and then wean it on to fish. So we were back again to the tube, but by now we had a better idea of what quantity we should give and what fat content to add.

As usual we sterilized the equipment. We gave the milk lukewarm in six feeds a day; the pup put up quite a struggle when first we tried to get the tube down, but we succeeded and after five days we decided to put it on to fish.

Sammy slept on a bed of straw, which was cleaned out regularly (the straw was to prove fatal later on).

This time for weaning the pup on to fish, I held him while Mary dropped the fish into his mouth. It stuck to his top jaw, and we had to use a smooth stick to push it to the back of his throat to get him to swallow. However, after a few unsuccessful tries, he took to his new diet, and finally fed without the need for a stick.

After that we took a large bowl of sprats to him four times a day. Mary would hold the bowl and hand a sprat at a time to me and I would give them to him. Occasionally he would throw one out on to the straw, and if I picked it up, he would snap at my hand, so I let it lie there. As soon as we came to the last few sprats, we had to be ready to get out of the door quickly for, as with Sampson, Sammy went for one when he saw there was no more food going.

I'll never forget one particular evening; as the nights were dark we used what was supposed to be a hurricane lamp to light our way. The wind was howling, and as we went down the hill to the chalet we covered the lamp, but still the flame kept on flickering. We arrived at the door of the chalet, Mary holding the bowl of fish and I the lamp. When we opened it the light went out. Sammy snarled, and we heard his flip-flop coming towards us; not being able to see him, we made the quickest ever get-away, but still the flip-flop kept following us. I hurriedly relit the lamp and went round the other side of the chalet, taking a few sprats with me. Now I saw Sammy and came up to him and showed him the fish, he turned round and followed me back into the chalet. He still dropped some fish, usually those whose heads were torn. Very often he'd later pick them up and eat them complete with pieces of straw. He did so now, but we didn't think much about it at the time.

Soon came another call; a seal had been found in very bad condition, those who had discovered it said they would bring it over to us. When they arrived, they told us that they had put it in the boot of the car, and after travelling about a mile, one of them noticed that the boot was open, and stopping the car they discovered that the seal had gone, so they went back and after half a mile they found it on the grass, well off the road and no worse for its adventure. This time they put it inside the car.

I helped them to get it out and we put it into the caravan. It was a male pup, about three to four weeks old, obviously it had congestion of the lungs, and it too was very thin. Its head was beautiful, but its eyes were not bright, for they were thick with a yellow secretion. I bathed the pup's eyes and put some lotion into the lids. We called him Simon and we let him rest for a few hours.

We decided to wean Simon on sprats, since they were smaller and easier to swallow. He did not snap at me when I caught hold of him to give him the fish, and I opened his jaws, using gloves. At first he seemed to have some difficulty in swallowing, but he soon got the idea and we were able to give him a good feed but we took care not to overload him.

Later, we had him injected, and obtained an antibiotic which we gave him twice a day in the fish. As sprats are very small, which makes it difficult to put powder into them, we tried him with a small mackerel. First we gave it without the powder, pushing the fish down his throat; he took it fairly easily. The next fish I split and put the required dose inside it. Simon didn't notice the difference.

I did not like keeping him in the caravan, so I bought a second-hand wooden shed and placed it on the lawn at the side of the garage. The first few days he ate well, but his breathing was very bad and I took care to keep him very warm. I also made a wire netting run in front of the shed, so that, later on, I could bring him out into the sunshine.

In the meantime, Sammy seemed to be improving, but he did not put on much weight. The vet gave me another powder to put into his fish and we added cod liver oil and some extra vitamins. Unfortunately, owing to our lack of experience we did not recognize the signs or symptoms of what was happening to Sammy. Worried about him, we took him up to the garden and put him near Simon.

It was now that I started to build my first small pool. Since I had found that the seals were best weaned on to fish as soon as possible, I thought I must teach them to eat fish in the water, as they would have to do at sea.

The following day Sammy died. The post-mortem showed straw in his stomach blocking the intestines. We learned by these sad experiences and they were never repeated. As a result a good many of the pups we rescued are now safely in the ocean.

It was a sad blow to us losing Sammy, we had given him a lot of our time. When a fish fell on the straw he must have picked it up after we had left him and each time he had eaten the straw as well. Future seals were never fed near straw, and any fish that fell on the floor we removed before we left.

CHAPTER FOUR

Simon

SIMON, whose ribs we could see when he came to us, was fattening slowly. I let him out on a warm day into the netting run, he enjoyed this. I went in to dinner, Simon was sleeping outside, all looked well. Twenty minutes later there was a knock at the door, then a man came in and told us that Simon was fifty yards up the road outside the local pub. I went running up, there he was having a look round, perhaps he was feeling thirsty and wanted a quick one?

He let me pick him up. He was getting quite heavy now. I took him back to his shed and secured the netting he had wriggled under so there would be no fear of his disappearing again. As I was fixing the netting he looked sadly at me. I think he wanted some attention so I started playing with him, tickling him on the tummy, and under his chin.

I carried on with the pool, occasionally spending a little time with Simon. It looked as if his lungs were permanently damaged and he would never go back to the sea, so the quicker I could finish the pool the better for him. I worked day and night, and when I had put the final touches to it, I bought some wooden fencing to put round the pool, for two reasons: to keep Simon in, and also to act as a shield against the winds.

When the concrete had hardened off, I filled it with water. We were excited to know what Simon's reactions would be when he went into the pool for the first time. I had made a gradual slope to it, only six feet from where Simon was resting.

Now I removed the netting with which I had covered

the pool while the concrete was still soft. Simon just lay where he was for a few minutes so I went to the other end of the pool and waited. Then he lifted himself up and started to come towards me. It had been raining, the concrete at the side of the pool was smooth and wet, he began to slide and splash – he was in the pool. The water level was six inches below the top of the concrete; Simon tried to get out but kept sliding back. Now I threw a coloured ball into the pool, he swam towards it, then pounced on it and the ball flew up into the air. He did this a few times then, lying on his back, he came to the side. I tickled his tummy, he loved this and kept on smacking his chest asking for more. I was still not sure about how he might use his teeth. I stroked his head but as I did so he tried to grab hold of my hand. I let him hold my sleeve, it soon started to unravel, so I put on a glove and took the risk of his teeth biting through it. He pressed hard at first, I think, in excitement, then he caught hold of the finger-tips of the glove and pulled it off. I thought, here goes, and put my hand to his nose, he opened his mouth, I could see his teeth glistening; he caught hold of my hand and closed his mouth, not hard, but just hard enough to leave teeth marks. I kept talking to him and gradually he eased his teeth off. After that I could tickle his tongue and do as I wished.

He had a wonderful temperament. He would start to show off by nosing the ball, then he would pounce out of the pool; when on land he would wriggle his body, move his head backwards and forwards teasing me and trying to get me to play. I would kneel and pretend I was taking no notice of him; then he would crawl towards me and nudge my arm with his nose. Still I would take no notice and he would scratch me with his flipper, after this, if he got no response, he would roll on his back making grunting noises. At this point

I *had* to play with him. After a while I would push him into the pool, but no sooner was he in than he would pop out again. I noticed that his breathing was very hard, whenever he chased me, he gasped, like a person with a heavy cold. He was now well over 120 pounds and when he rested his weight on my feet I could not move. If I tried, he would grab hold of my trousers, and I'd be less one leg of my trousers if I continued to endeavour to escape. Simon could move quite fast over a short distance, his body lifting up and slapping down on the concrete.

Many people came to see the Sanctuary. This gave Simon the chance to show off when I was not there. Sometimes, when he was fast asleep under water, only very occasionally coming up for air, people would arrive at my door saying he was dead.

I put a hose-pipe connected to a tap into the pool, Simon would wrap this hose around himself thus supporting his body and would often go to sleep in this position.

I was still working just outside the fencing of the pool and sometimes Simon, using his flippers, would loosen two strips of wood, it was interlacing fencing, and pop his head through to see what was going on.

That winter three more pups were washed in and another two found dead. They were successfully weaned on to fish and returned to the sea. One was with us only four days, another a week, the third eleven days.

During the summer months I could not give Simon very much of my time, but a neighbour who took an interest in the work we were doing and who was attracted to Simon, used to come and play with him, and if I could not go to the pool when it was his feeding time, she would feed him for me. Of course, I came along as soon as I could; when I reached the gate out he would jump from the pool, and behave like a dog would if you'd been out for the day, flapping across towards me, putting his teeth around my jacket, then trying to climb on to my back. If I had my cap on that would be the first thing he would go for, taking it off my head. He was now devoted to me and if he had had his way he would never have let me out of the pen.

On one occasion Mary and I had to leave the Sanctuary for a few days owing to her mother's illness. We arranged for a neighbour to feed the seals and care for them, and every night I telephoned to hear how things were doing. When we returned Simon behaved just as some dogs do when their owners have left them for a while. As soon as I reached the car park I looked down at him and shouted to him, usually when he heard my voice he would be out of the pool in a jiffy but now he just looked reproachfully at me. I went down to the pool to talk to him but he ignored me and when I tried to stroke his head he went off to the other side of the pool. It took me nearly ten days to get him out of his fit of sulks and back to our normal relationship.

In July, Simon started having lung congestion again. I put an antibiotic into his fish but the following day he was worse, so I rang the vet to give him an injection and I emptied the pool to make it easier to get hold of him. But, although we were great friends, Simon at once sensed that something untoward was going on as soon as the vet jumped into the pool. I held him tight around the neck and told the vet to inject him quickly. Simon was not at all pleased about this, so, after it was over, I let him rest in the empty pool to give him a chance to recover from what must have been an embarrassing situation for him.

That night I myself jumped into the pool and talked

to him for half an hour. By then he had got over the injection, so we were friends again. I left the pool empty that night, letting him rest, as the weather was good, and in a few days Simon was back to normal. He ate well, in fact, by September, I was worried about his size; the weight he was carrying was considerable for a weak-chested seal and I gradually cut down his food, even so, he was eating fourteen pounds of fish a day.

Simon was a great favourite especially with children who, before going home, would come to say 'Goodbye' and 'See you next year, Simon, take care of yourself'. Even old people came on the last day of their holiday to bid him farewell.

I had just closed the shop for the end of the season when the first pup of the year came in. He was the usual creamy white bundle. We used the wooden shed that Simon had at first been in. We were more experienced now in handling and feeding, so it seemed much easier.

We kept the pup clear of Simon as we thought it would not be fair to let them get used to each other when they would have to be parted in a few weeks' time, supposing that the pup was returned to the sea. This pup was no problem, but the one we got two weeks later died the following morning, though he had been injected for pneumonia.

He may have been on the beach for days and in that case was past the stage at which any treatment could do him any good. Most of those we found that winter were either injured or had been separated from their mothers by gales. All the same, it was a successful winter for us; only that one pup died, and not through our fault as we had received him too late.

CHAPTER FIVE

Sally – a victim of the Torrey Canyon

DURING Easter week, 1967, the Torrey Canyon went adrift on the Seven Stone Rocks off Land's End. Attempts were made to pull her off, but gradually she started to break in two and the Government decided to blow her up as there were thousands of gallons of oil aboard and already slicks of oil had reached our beaches. The sea birds were hit hard: they were coming in on nearly every beach in Cornwall. Centres were set up for cleaning them. At first we started attending to them on our own beaches, but owing to the numbers involved and also to the considerable time we were spending there in collecting them, it seemed better to take them to the nearest cleaning centre, which was at Perranporth. We were out all day and all night. In the evening we switched the headlights of the car on so that they lit up the beach, and we collected many birds this way. We also got many a soaking for it was easy to trip over stones or large rocks and go sprawling into the sea. The beaches were thick with oil: it was a horrible feeling just walking through it while the poor birds were suffocating in it.

There was an uproar for holidaymakers would soon be arriving and nearly every beach was plastered with oil. As fast as we moved the oil off by filling trucks and dumping it into an old mine shaft near the beach, more oil come in and the coves were as bad as ever. Later, troops, the Fire Brigades and members of the Civil Defence Force all worked to clear the beaches with detergents and besides that, many people helped, giving up their time to try to save the birds. Nevertheless, thousands died.

A rough calculation suggests that as a result of the disaster, over ten thousand birds died. Despite all the work involved in catching and cleaning them, very few survived. Indeed I think it is a question whether badly oiled birds should not be destroyed immediately, instead of being left to suffer over a prolonged period.

Birds covered in oil cannot fly and therefore battle with the sea; they become weak, their feathers are soaked and they suffer from shock. When, finally, they reach the rocks or the beach, we add to their strain by chasing them, even though we do it in good faith, trying to save their lives. After this they are placed in a dark box, which adds to their distress. Then they are carted around, other birds being put in with them, and sometimes fighting with them. When they arrive at their destination, their feathers are rubbed with solutions intended to clean off the oil; finally they are placed in front of fan heaters. All this is unnatural, and, I think, in most cases, proves too much for the bird. The dilemma is that if we do not try and save some, then we may lose our entire population of sea birds. What is the answer? Should we try to save all of them, thereby perhaps causing further suffering to many, or try to save only the lightly oiled birds and put the rest out of their misery, or should we destroy all the oiled birds? The only satisfactory answer would, I think, be a dry powder which would destroy the oil without the birds having to be washed; such a powder would need to be applied as soon as the victim is caught, and I believe a lot is being done to try to invent something of the sort.

Every time we went on rescue expeditions it seemed that the winds howled and rain poured down! We covered every possible place from Perranporth to Godrevy. No one wants to see another *Torrey Canyon,* but who can tell? The tankers now being built are bigger; errors can be made by the skipper or crew.

While this was going on, I was also getting phone calls about seals in difficulty, and I think I covered the whole of Cornwall in answering them. Most of the seals were, in fact, all right, but as they were swimming in slicks of oil, watch had to be kept in case their eyes were affected. To check the situation properly we formed a cliff party: ropes and harnesses were loaded into a jeep and a search of the coastline was organized. There were five of us in all – none too many – since one or two of us might need to be lowered down the cliffs for a seal rescue operation. We had many a frightening experience whilst being lowered, but all of us were good men on the ropes so the only horrid thought was 'What if the rope were to break?' We did however get one seal who had been blinded by the oil. She was a female, we called her Sally. She was plastered in oil and it took some time to remove it. There was not much we could do about her eyes, one was completely gone, and the other had only blurred vision.

We put Sally into the shed near the pool, washed off most of the oil, treated the eyes and let her settle down. The poor little thing was going to need a lot of care and comfort.

Sally was at the weaning age so we put her straight on to fish. Probably because her sight was poor she was very nervous and attacked anything that touched her, including me. We had to be very careful when handling her. I think she knew we were trying to help her, but she wasn't having any fuss, so, as soon as she was fed, out we got. The blind eye was healing but the other had a white mist over it and still looked sore. She also had a nasty scar on her side, but we were assured this would soon heal.

Simon had started throwing his head back when he

was swimming. I thought perhaps a fish bone had stuck in his throat and would probably loosen itself in time. But as he kept on doing this for days, I began to wonder whether he had some sort of infection in his gullet. My vet called to examine him and we decided to X-ray him. The nearest vet who had the right equipment was at Exeter, so we contacted his office and made the necessary arrangements. The pool was emptied and my own vet came to inject Simon hoping to put him to sleep. The maximum dose was given so that by the time the X-ray unit arrived he should have been asleep. But, in fact, when they came Simon was only a little dozey. We tried to put the X-ray unit by him but he snarled and snapped. So that he should not get too distressed I went down to reassure him, and gently held his head.

The unit was then brought in and slowly placed over him. He did not like the machine and I had difficulty in holding him. However, in the end, a few X-rays were taken. It took nearly a week before we got the result, which was negative. There was no fish bone stuck in his throat. We therefore treated him as for an infection. After two weeks he stopped throwing his head back, and that was another worry over.

Sally, in the meantime, was slowly improving as to health, but not as to temperament. She would still snap when I tried to touch her. She didn't mind people getting close to her, but she did not like anyone touching her.

I decided to build another pool at the side of the existing one. This would enable me to wean future pups in a pool of their own, away from Simon. I made a start but was not left alone for long. First Simon put his head in the hole I had dug, then he grabbed my sleeve and pulled me away.

Very soon, I let Sally into the netting run. This made Simon inquisitive; out he came to sniff at her. She was not very keen on this proceeding, gave a cry and shook her flipper. Simon, being good natured, did not take too much notice of her hostile attitude and just slithered back into his pool, occasionally coming up to the side, and popping his head up to see what she was up to.

When I had finished the new pool, I put some netting to divide the two and then let Sally into hers.

In the water Sally looked a pathetic figure. Her one eye completely gone, the other one white, but all the same, she had a sweet face. After a few days I decided to take down the netting between the two pools and see how Simon and Sally would react to each other.

At first neither of the seals attempted to go close to the other. Sally's pool was a little higher than Simon's, with a smooth concrete slope down. It had been raining and all the surrounds were wet. As I went in, Sally jumped out of her pool, and in doing so slid down the slope into Simon's pool.

There was quite a scuffle, mainly I think because Simon had been caught unawares. Then they sniffed at each other and afterwards Sally swam away, with Simon sniffing her back. I dared not leave them yet for if they didn't get on, Sally would have to go back to her own pool. I need not have worried, they got on beautifully. Simon kept his flippers round her: he thought she was wonderful. His face was mottled brown now, at first it had been mottled dark grey.

It was spring, the sun was shining, and both seals spent most of their time basking in the sun. I then noticed light specks coming out on the bald patches of Simon's skin (he had lost most of his fur some time earlier), and in a few days I realized he was growing a new coat. It felt soft and silky and it seemed no time at all before he had a coat as beautiful as the one he had when he was washed in. I think Sally fancied him all the more now that he no longer looked like a tramp.

On one occasion when I went to feed them, Simon was missing. I searched everywhere, there was no opening in the fencing; he could not have got out, I couldn't understand it. Then I heard a sneeze behind the shed which was against the back wall. (The shed was raised up on blocks and planks as it had a wooden floor.) There was a drop behind it and there was Simon, wedged between the wall and the bottom of the shed. He was not hurt, but, fat as he was, he couldn't move. He looked at me as if he were saying, 'Aren't I a right nit'. I could not move him an inch, finally I had to cut a hole in the back of the shed near his head and let him get out that way.

I had a rockery built up in stages, each stage about three feet high with a platform between each layer. One day I found Simon on the second layer stretched out and enjoying the sun. I held a fish to him to entice him down, but he didn't respond. He was scared stiff. He had got there by climbing, but, with his figure, getting down was going to be a problem. I climbed up and put my back just beneath him hoping he might use it as a stepping stone, but he was evidently too afraid of falling, so I fetched two long thick planks and put them side by side, and left him for a while hoping he would slide down them, but when I got back he was still there. Finally, I held a fish in front of his nose and edged it forward slowly, then he moved, placing his flippers on to the planks; he was still not sure whether they were safe or not, but eventually he edged on to them and slowly but surely made his way down. He never attempted to get up on the rockery again.

Having heard how intelligent the grey seals were, I thought I would try to teach Simon a few tricks. I didn't want him to do anything he did not already do but I wanted to see if he would do these things, such as twisting in the water, nosing the ball from one end of the pool to the other and splashing the water with his flipper when he wanted more fish, at my suggestion. So, when it was feeding time I used the word *twist*, gesturing with my hand to show him what I meant. At first he splashed, then suddenly he twisted. I threw in a fish and tried again. Once more when I said *twist*, and used my hand giving a circular motion, he twisted.

Next I threw the ball to the other end of the pool and used the word *fetch* pointing with my finger. He went after the ball all right, nosed it and came back, so I rewarded him. At least he had gone to the ball. Each time I threw the ball he went after it, but only to toss it out of the water. I thought that was enough for one day, but not Simon, he first twisted, then he waved his flipper in the air, as if trying to splash the water, and, as a last resort, he went to the ball which was floating and flipped it up in the air. I had no fish left so all I could do was to tap him on the head and call him a good boy.

Sally, unfortunately, couldn't take part in any of this; all she did was sit up in the water on one side of the pool. She did not have a lot to do with me, but Simon saw to it that she was never lonely. Whenever he had finished his play with me, he'd chase Sally round the pool and out she would jump followed by Simon's huge bulk right on top of her. Sometimes they made a noise and their teeth seemed to be biting, but they never hurt each other, and all this was only by way of play. Often they would chase each other around, and usually it ended by one or both of them falling into the water. This provided much amusement for the many visitors who came to the Sanctuary. To explain to them why the seals were there and why the Sanctuary existed, we put up posters telling them how the pups had been washed in and the illnesses they had suffered from. All the same I was asked a lot of questions. 'How can a seal catch pneumonia?' or 'Won't they die out of water?' Many people showed great interest in Sally and Simon and some even sent Christmas cards addressed to them. The amount of pleasure that these seals were giving to the old and young alike surprised me; some people had to be dragged away from the Sanctuary as they wanted to stay there all day. A number of them said how much they would love to have seals in their garden, little realizing the work involved.

By September, Simon would, on command, ring the bell, fetch the ball, flipping it into my hand, do the twist, splash the water and smack his tummy. These acts he performed without being rewarded with fish.

Every time he saw me he would tease me in the way a puppy teases his master, running to and fro, sometimes biting at my trousers. Considering that their new home was small, both Sally and Simon were happy, but I was already planning how to make a much bigger pool out of the two small ones.

At the end of September, another newcomer arrived at the Sanctuary. Creamy white coat, bright eyes and not in bad condition. But to make sure we had it injected with an antibiotic, and then tried it with a bowl of milk to see if it would lap – it wouldn't. I decided to buy a calf feeder and find out whether this would be useful for weaning the babies; I wanted anything that would make the process more natural for the baby itself. The calf feeder consisted of an aluminium container with a large screw-on teat at the base. It had a lid and a handle, also a fixture for hanging it on a wall. We kept the milk in the jug, thinking that if the pup would not suck at the teat, we could still use the old tube method.

Now, we put about a quarter of a pint into the container and stuck the teat into the baby's mouth. At first, it kept moving its head and the teat kept coming out, so I held the teat, put it into the pup's mouth and gave it a squeeze, thus letting out a little milk. It bit on the teat but each time it released the hold I gave another squeeze, while keeping its head up high, at last I heard a gulp – it was swallowing the milk. Finally all the milk had gone. At least it had fed without the tube, but the queer part of it was that after moving the feeding apparatus, the pup suckled at my hand. The suction was very strong and why it did not do this on the teat I did not know.

As it was a sunny day we gave the next feed outside. Simon popped up to see who we had now. I started feeding the pup when Simon came alongside me; first he put his head across my lap to sniff at the baby, then, as I was using both my hands on the feeder, he caught hold of my sleeve and began to pull it away. I held the feeder tight in one hand and stroked him with the other till he calmed down. As soon as I had finished feeding the pup, Simon moved in closer to inspect the new arrival more thoroughly. This time I let him have his way, but I did not wish them to get too well acquainted, as this pup would be ready for the open sea in a few weeks' time and it would not be fair to let them make friends only to be parted.

The feeder had done its job. Weaning time was on us. We decided to do it outside so that there would be no fear of straw getting on to the fish. We used small mackerel and had no difficulty in pushing them down. Gradually the pup began to take the fish out of my hand, so, using Sally's pool, I threw fish into it hoping he would pick it up. If he did, he would be ready for the sea.

Winstone, as we had named it, made no attempt to go for the fish, so I lowered the next one down to the water beneath his head, and kept on doing this till finally he took a fish when it was immersed in the water. The following day he was picking them up at will and we made up our minds that after another day like this he could have his freedom and perhaps join his family. He had put on a considerable amount of weight and carrying him to the sea would not be easy.

We decided to go early, about six in the morning, hoping there would be no dogs or people to distract Winstone. It is never pleasant taking a seal back to the beaches, one gets attached to any animal and most of these pups had never known their mother and so had come to feel reassurance in our presence. One is left wondering uncomfortably what happens when they meet up with their own kind. It was only later that we proved that they do meet and join other seals, whether these may be their own family or not.

We carried Winstone down to the water's edge and gave him a kiss and a word of good luck. It was sad: that great big ocean in front of us and the little pup which we were leaving alone to fend for himself. However, it would have been wrong to have kept him, Winstone had not been sick, just orphaned. And all we had done was to prepare him for the life he was meant to live.

We put him carefully on the sands and smacked his bottom gently. He made his way into the water, sometimes he gave us a side glance, then he dived under the water and was gone. We watched and glimpsed a little head out in deep water. He made his way around the point and that was the last we saw of Winstone.

CHAPTER SIX

Intruders

SIMON and Sally were both eating well. Now that Winstone had gone I thought I could spend more time with them but this was not to be, for we soon had another two arrivals. Both were only a few days old and both were males. We called them Flipper and Moses – Moses because we found the baby covered in seaweed.

We divided the shed into two, and put one in each compartment; neither would lap or suckle at the teat, so out came the tube again, this time twice the quantity of milk to be warmed and mixed. But now we found force feeding was fairly easy. Sometimes a pup would play up, but neither Flipper nor Moses caused us too much trouble. We gave four feeds a day. But the pups did not put on weight as they would have had they been suckled by their mothers. However, after losing our first pup by overdoing the fat content of its diet, we had no intention of letting this happen again. So long as the pups kept healthy and gained a little weight we felt it was satisfactory, particularly as we usually weaned them fairly early and they did not have to battle against the sea. In their natural environment they would have had to do just that and would have used up their extra fat while learning to feed themselves.

We had snow that winter, only a few inches, but it was bitter cold. Simon and Sally did not seem to mind it, and Simon dug his nose into the snow and flipped it into the air as he moved along. He would start sliding and, not being able to stop himself, would end up in the pool. This was great fun for him so he repeated it several times over.

Having four seals kept us busy, but we enjoyed every minute of it. Very often I would be covered in muck or milk and people who came to the Sanctuary did not have to ask me if I were the keeper; for one thing they could smell me. Seals have a smell of their own, and when they bite one's clothes their saliva, which has a special odour, gets all over one's sleeves.

Flipper and Moses had settled in, took their milk and then enjoyed a good sleep; their little heads would pop up when Simon barked at Sally. The two new pups brought many people to the Sanctuary, who each day enquired how the babies were getting on.

While they have their creamy white coats, seal pups look quite plump, but when they moult and the grey fur shows, they look much thinner and when in the water look smaller still. I used Sally's pool for Flipper and Moses's first swim, which was fun for them and also cleaned them. As they could not get out on their own, I'd grab hold of their front flippers and give a heave till they slid up. But after a few pulls they managed it for themselves.

Weaning was very easy with Flipper, but with Moses it was much more difficult. We could force feed him, but he would not take fish on his own. Already Flipper was taking fish in the water, so he'd soon be ready for returning to the sea.

That night I locked up, shouted goodnight to all of them, and went to bed. The following morning I put their fish in the bucket and went over to the Sanctuary. The gate was wide open, Sally and Simon were in the pool, but there was no sign of Flipper and Moses. The lock had been forced open and the two babies were missing. I made a thorough search around the shed, then around the bungalow, but found no sign of them.

I ran to tell Mary; together we made another search. We went all round the car park behind the bungalow, along the cliff road and finally down the slipway to the beach. There was a track leading down the beach to the sea, we followed it but saw nothing. We did not know whether Flipper or Moses had made the track and we couldn't understand why there was only one track. Had the vandals taken the other seal away? We searched all round the beach, then we went up to the boat pen and looked under all the boats, there was no sign of any track or of a pup. Then I went to a small building in which we kept a winch for pulling the boats up the beach. At first I could not see a thing, then I heard a faint movement. I went in, looked behind some boxes in the corner and there was Moses. There was blood all round his head and a cut on his body and some on his flippers. He must have been frightened to death and at first he snarled at me.

I kept talking to him and when he realized who I was I was able to pick him up and carry him back to the Sanctuary. On checking his injuries I found that his cuts were not too bad, but whatever had happened during the early hours of that morning had shaken the poor pup very badly.

After settling Moses, we rushed back down to the beach to look for Flipper. We did not feel too bad about him for he had been weaned and should now be able to fend for himself, but we were worried in case he too had been injured. We walked along the water's edge for what seemed hours, then we decided to go on top of the cliff, where we would have a good view over quite a large area. As we were walking we kept on shouting 'Flipper', knowing only too well that the roar of the sea drowned our voices. We felt mad with the world: what sort of person would do such a thing and what satisfaction could they get out of it? A further search was fruitless, but some friends of ours kept watch along the beach in case Flipper might be in difficulty.

The following morning we had another search. At first there was no sign of the pup; then we saw a little head pop up in the water about thirty yards out to sea. We shouted 'Flipper' as loud as we could. He raised himself in the water, but whether or not he heard us I do not know.

We stayed for about an hour, while the pup swam along parallel to the beach. As it was lunch time we then went home, but from our windows we were able to look down on the sea and beach. Just as we had finished our lunch, I looked and saw Flipper roll out of a wave on to the beach. I ran down, but before I got down to the beach he was back in the water.

It looked as if he might be all right though not yet quite settled in his new home. Only one other time did he come on to the beach and again it was only for a few seconds. Often during that summer he swam in the bay and played with the lads on their Malibu or surfboards. At least it was nice to know that he had made it safely, but that was no thanks to the vandals.

A week later we put Moses back to the sea, but we never found out if he met up with Flipper, or whether he went his own way and sought his pleasures further down the coastline.

After this encounter we double-locked the gate to the Sanctuary and had no further break-ins.

The following Tuesday evening about five, we had a telephone call telling us of a seal washed in, and badly injured. Darkness was coming down, the winds were blowing gale force and there was heavy rain. It was not too pleasant having to leave a nice warm fire, but we didn't really mind. In any case the poor little thing must be suffering a lot more than we would have to suffer just getting wet and cold.

I started the car, and we gathered all our gear together: wellingtons, thick coats, scarves, gloves and a torch that showed no more light than a match would. We had nearly thirty miles to go so there was no time to waste, particularly as we couldn't drive very fast since the heavy rain made it difficult to see and the roads were slippery. First, we made for the house of the people who had telephoned to us. As it was so rough and wet we could not ask them to accompany us to the spot where they had seen the seal. But they gave us precise directions, and we started on our trek across the beach. Because we were in a hurry I kept forgetting about Mary and as the light from the torch was very poor, she was calling out that she could not see where she was. I had to go back to collect her and then keep shining the torch in front of her as we tried to make our way to our destination.

Because I was shining the torch behind me to help Mary I fell over a rock and went sprawling into a pool of water but, looking up at Mary, I saw that I couldn't be any wetter than she was. Water was dripping all down her face, her trousers were sopping; we looked like a pair of drowned rats. What had we got ourselves into, out on a night like this fetching a sick animal? A few years ago we would never have dreamed that such things could happen to us.

We had one or two further stumbles before we reached the place where the pup was said to be. With the wind blowing into our faces and the torch only giving a glimmer of light, it was like looking for a needle in a haystack. We listened for any noise or movement; suddenly we heard a sneeze just in front of us behind a rock. We edged forward and could see the seal turning towards us, ready to protect itself against any menace that might be approaching. As we got closer, we could see it was very thin. It had lost its baby coat and had a nasty gash down its back. Obviously here was another pup that had not learned the art of catching fish for itself. When left on its own, it had lived on the fat its mother had provided it with and had then gradually become weak from starvation, also it had been pounded on the rocks and been badly injured. I examined it, and decided it must be got home as quickly as possible, if we were to save its life.

I did not think twice about its biting me and grabbed hold of it quickly, but carefully, so as not to cause it any pain from its cut. I got both arms around the pup's body; Mary shone what little light there was left in the torch and we started back for the car. How I managed to keep going without having to put the pup down on the sands, I don't know. It seemed to get heavier every yard I went, till I was gasping for breath and water was running down my face into my mouth. I was hot and cold both at the same time. It was a relief when we reached the car, but I had no strength left to lift the seal into it. Mary came to my help and between us we got the pup on to the seat, and made it comfortable. Then I had to wait ten minutes to recover from my cold shivery feeling, and also to get my breath back. We threw our wet jackets into the boot but our trousers were drenched and we had a very unpleasant journey.

Before leaving, we notified the people that we had found the pup and then drove home as quickly as possible.

On arrival, we carried the seal into the shed, put on the infra-red lamp to give a little warmth and then we dried ourselves off. Once changed and warmed

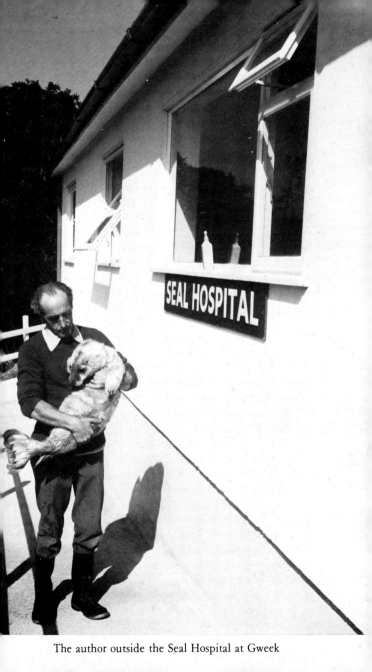

The author outside the Seal Hospital at Gweek

Rescue of a seal pup

Pup gaining confidence after being fed by tube

Simon I looking for company . . .

. . . and wanting to play

Simon I being X-rayed

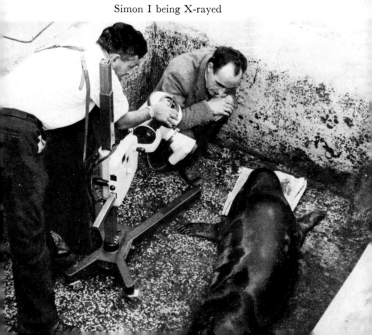

Sally II meets new pup

through we made a check on the pup. The cut was a bad one, but seals' wounds usually heal fairly quickly. The main thing was to stop any infection. We had it injected and obtained powder to rub into the cut to dry and treat the wound. Unfortunately, in spite of all our efforts, it died four days later.

CHAPTER SEVEN

A double tragedy

AT feeding time, there was never any fighting over the fish, Simon and Sally waited patiently for their own share, Simon of course doing his usual tricks. The way they had taken to each other offered good prospects for breeding later on, and we were looking forward to this.

During the summer we had more trouble from people throwing rubbish into the pool: polythene bags, ice cream sticks, dirty old corks and rubber. Each day I had to collect rubbish from the pool and its surroundings although there were many signs asking people not to feed the seals or throw anything into the pool.

I put up more netting to try to stop them, but still they found a way. Even though I was busy in my shop, I made a point of having a peep into the Sanctuary every few hours to check how the visitors were behaving. It had become a great attraction and seemed to give a lot of pleasure to old and young alike. Many of them had never seen a seal before, and few had been so close to one. Simon would come right up to the netting, he loved human beings, but Sally preferred the safety of the pool.

Sometimes there were fifty or more people in the Sanctuary but if I went amongst them and my head showed up, Simon would be out of the water in a jiffy; this proves that seals can recognize one by sight as well as by sound.

The summer season visitors made it difficult for me to clean the pools. I had to do it after midnight when the business was closed.

The hose-pipe I had fixed up for the seals had often caused laughter among the holidaymakers. They'd be enjoying watching the seals play when suddenly Simon would catch hold of the end of the hose-pipe and squirt water over them.

I had been thinking of a way to make one large pool out of the two small ones. When my shop was empty, I drew rough sketches of what might be possible. I had four ship's wheels and I planned to put them into a wall of concrete blocks faced up with stone. Finally, I telephoned to one of my suppliers of concrete slabs and arranged for him to make a hundred white and red two-foot square slabs. They would be ready, and hardened off, by the time I needed them.

I also planned to take down my wooden garage and build one of concrete blocks with a room at the end as a sick-bay for seals. I also wanted another sick-room inside the Sanctuary; this would be used for hardening the pups off after being in the heated room attached to the garage. When all this had been done I should be able to cope with quite a lot of babies if several arrived at once, as sometimes happened.

At the end of September I put my plan into operation. I started to dig out the centre piece between the two pools, then, putting up netting close to the edge of the pool, I emptied the top one; after this, I was able to knock through which gave me about sixteen feet to the edge of the pools Simon and Sally were in. So, with their existing pool I could make a pool about twenty-four feet long by twelve feet wide, not extra large, but the largest I could organize at the moment.

When all the earth and rubble were removed, I put in a thick concrete base and then started to build the walls. My plan was to get Simon and Sally over to the part I had built and put netting across to keep them away from the new section I still had to make. The seals would come to no harm if they had to remain on land for a day or two.

When my work was done, the pool was filled and out came Sally with Simon chasing her. The slabs were smooth so they went slithering along them and I think they were proud of their new home.

Our first new pup of the winter came in October, and between that date and December six were washed in. One was in very bad condition and died. The others were successfully weaned and put back to the sea. We were lucky that they were not seriously injured and were able to be weaned at an early age so that they returned to the sea before they had become too attached to us.

Now that the pool was much longer and wider, Simon had a lot more room to show off in; in the middle I had placed a pillar eighteen inches square with a water pipe built into the centre. On this I put a spray; Simon loved it and would leap into the air trying to catch the water, after which he would roll beneath the spray as though he were taking a shower bath.

I had a few lifebelts in the house so I dropped one into the water, another I hung up on to a bar, the belt just resting on the top of the water. First Simon popped his head up through the lifebelt, then he put his flipper over the sides of the belt, using it like a child would in the sea. After that he would try and get on top of the belt, usually falling off at first, then, when his whole body was on, he'd roll on to his back and use it as a float; moving his rear flippers slowly he would travel up and down the pool as if in a boat.

When he got bored with this performance he would swim down to the other lifebelt, put his head and flippers through it and support his whole body in it; it acted just like a see-saw balanced in the middle; on this he would go to sleep. When Simon didn't use the belt, Sally would get inside it.

In May of that year, 1968, Simon had another attack of lung trouble; he looked really rough. I emptied the pool and put a large dose of antibiotic in a fish, but he wouldn't eat it. I rang my vet and asked him to come quickly and give Simon an injection, for this was the only other way of treating his complaint.

Again I had to hurt Simon's pride by holding him tight while he was being injected. Although I could do anything with him when no one else was in the pool, as soon as the vet appeared he would snap at anyone. After a struggle, we managed to inject him, then he was left to rest and Sally came up to his side as if to comfort him. A few days later Simon seemed to be himself again and I filled the pool.

But by the first week in June he didn't look well. I treated him with antibiotics; they had no effect. I watched him closely during the next few days, in fact I couldn't settle to my work in my shop where people came in constantly to tell me how poorly Simon was looking.

One morning Simon wouldn't eat. I wasn't too worried about this as sometimes seals have a fasting period, but the trouble was that if he didn't eat I could not give him his medicine. About midday I went out to see him, he was gasping for breath and looked really sorry for himself. Having been with seals for so long I could now tell when they were off-colour and when their condition was serious, and I knew that Simon was very ill indeed.

I rang the vet; he came rushing out, but by the time he arrived Simon was in the water cuddling Sally gently. The vet was inclined to think that I had brought him out for nothing. He said Simon seemed all right and left.

A little later I went back to the pool. Simon came to

the side; he held his nose slightly out of the water and curling his body up squeezed his neck making it twice its normal size.

I called to him, but he paid no attention. Sally came over and sniffed him, but he took no notice. Then, after five minutes, he suddenly swam around the pool, returned to the same place and started to gasp for breath. He went under water, but in a few seconds was back up again, this time he was gasping still harder.

I held his head above water. He looked pleadingly at me as though trying to tell me something I couldn't understand. I spoke to him gently thinking this was only another attack, such as he had previously had and recovered from.

After a few minutes I released his head, but again he gasped and this time he went into a frenzy, then suddenly he seemed to be drowning. I quickly turned the stopcock to empty the pool. As the water slowly went down, Simon gave a sudden leap up into the water and made a dive back, which gave me a fright. I realized that he couldn't breathe, so I jumped into the pool – clothes and all – and held his head above water, but he didn't stir. He weighed about five hundredweight so I couldn't lift him. I waited until the water was out of the pool – it only took a few minutes to empty – but it seemed hours to me. Then I shook Simon and spoke to him, but he had no reaction. My first thought then was that he was dead. But that couldn't be. I gave him artificial respiration, with no result. I shouted to a vistor, asking him to ring my vet urgently and I told him what to say. But I soon realized that Simon had left us. He had been slowly dying for the last few hours, but he could not tell me. He had given Sally a cuddle, perhaps he knew that he was leaving her. Sally being practically blind, was so dependent on him, he was her protector; had he given her a last expression of affection to console her for what was to follow?

Simon was lying on his back. Sally came up to us and crawled up on to his tummy and scratched his body with her flippers. She pushed at him with her nose, scratching at the same time. As she was doing this I suddenly noticed tears falling from her eyes, not the normal watering that seals sometimes have, but real tears. (I must admit I too had tears in my eyes.) We had both lost a part of ourselves, the affection Simon had shown us during the six years he had been with us will never be forgotten; he had become part of our family.

Except for the occasional illness, he had lived a happy life and had given happiness to people from all over the country. We shall always remember Simon.

Sally kept edging up to his head. I'm sure she knew he was dead. She lay motionless by him, tears still coming from her eyes. Up to now I had never been able to fuss Sally, now I put my hand gently on her back and she didn't stir, so I stroked her, gradually moving my hand up to her head; under normal circumstances I wouldn't have dared attempt this. I think both of us must have been in a state of shock. I stroked her head as one would a dog's and gently she lifted it and put it across my arm. I couldn't believe it. She rested her head there crying like a baby.

I thought that Sally's grief must be much deeper than mine. Her blindness had made her nervous, but at least when Simon was there she was protected and comforted.

The vet arrived. He examined Simon to make sure that he was dead. He said that after seeing him earlier that day cuddling Sally he couldn't believe it and suggested a post-mortem. Although we had previously carried out occasional post-mortems on pups, I dreaded it. But the vet said that as Simon's health had been bad from the start, a sudden death was to be expected at some stage and now, by examining his lungs something must be learned to help future pups suffering from similar conditions. If knowledge could be gained by a post-mortem, then I must agree.

We decided it could be done in the shed at the side of the pool. Sally was still lying on Simon's body so we had to move her in order to get him out. By holding Simon's head and back flippers we tried to ease his body from under her, but she gripped him tightly and wasn't going to let him be taken away. Finally, as we slid the body carefully along the bottom of the pool, Sally became very upset and tried to bite the vet's hand.

We then fetched a large plastic sheet and rolled it underneath Simon and asked two strong men to give us a lift which they did willingly. As we tried to lift the body Sally hung on as hard as she could. It was pitiful to watch her, but Simon's corpse had in any case to be removed.

All this proves that animals have as deep feelings as we do. They love, feel and weep as any person would.

We carried Simon into the shed, closed the door and the vet prepared for the post-mortem. When we saw his lungs, we realized that nothing could have saved him. Over a period of years they had deteriorated, and one had collapsed. At least he had never suffered any pain. I arranged with a local farmer to fetch the body and have it buried in a friend's field nearby. When the tractor came to take it away people gathered around asking questions, but I was unable to answer them.

When I went back to the pool, Sally was lying there motionless. I spoke to her softly trying to comfort her. She let me stroke her. All this had taken an hour and a half and Mary knew nothing of what had happened.

I smoked a cigarette and drank a cup of tea to steady me, then I went to the restaurant. Mary was in the kitchen; she turned round. I expect my face looked a sight as she and the waitress asked if I were all right. 'Simon's dead,' I said. There was silence, which was finally broken by Mary asking a few questions: when, how, why?

I had a cup of tea and went back down to the shop only to be further upset many times by visitors who knew Simon and asked how he was. In the end I wrote out a statement saying that Simon had died that day and I posted it up in the Sanctuary. Hundreds of people came into the shop and expressed their sympathy.

The next few days were very trying; Sally wouldn't eat and though she accepted all the fuss I made of her, she showed no interest whatsoever. I had her injected for shock and gave her a booster, but still she wouldn't eat. Last thing at night I would go and talk to her, but she just stared into space.

A week after Simon's death, Sally too died. She had pined herself to death. A post-mortem showed a healthy body, but heartbreak has no physical signs. We placed her body next to Simon's and can only hope that in spirit they are together again.

It was a terrible shock to us to lose our two dear seals in one week but nothing could have saved them. Many seals have come my way, but never again one like Simon.

CHAPTER EIGHT

Sea-lions in the Sanctuary

IT was June and thousands of people came to visit the Sanctuary, only to find it empty. But it wasn't long before two sea-lions, suffering from worms in the lungs and intestines arrived; they came from a zoo. It seemed that, by now, people had come to think of me as someone who was always ready to care for sick or injured animals. After handling grey seals they seemed very queer to us. They were so lively and the speed at which they could move was something to which we were not accustomed. A sea-lion walks on its front and back flippers, moving in leaps and bounds; when walking slowly it has a sort of wobble, like a woman walking in a very tight skirt. They also differ in many other ways from the Atlantic grey seal.

The sea-lion is called the eared seal, having ear lobes about an inch and a half long, whereas the grey seal has only two small openings hardly noticeable unless they are listening to something, when two small horns project out of the orifices.

The sea-lion's front flippers are long, black and rubbery and have no gripping parts; they are used mainly for swimming and, of course, for walking on. It is these which give the sea-lion a sweeping effect and ensure its speed in the water. The back flippers, which can be tucked underneath the body when on land, are used as rudders in the sea. Their flippers are not at all like those of the grey seals which are short, thick, hand-like flippers with long thick nails in front, it is with these 'gripping hands' that seals climb rocks and pull their great bulk out of the water; the back flippers which they use for propulsion in the sea have no use on land.

Sea-lions look more streamlined, partly because they have a smaller head. The grey seal has a thick body which tapers to a point at the back flippers, its head is much broader than that of the sea-lion, indeed it is often called 'Horse Head'. Because the sea-lion has a graceful appearance and movement it is very popular in circuses, and they are trained to do many tricks.

We named our sea-lions Judy and Kim. They found the pool empty of water. We had our reasons for this; they had come in large wire crates, so now we lowered them down into the pool and opened the doors. Out came Judy, giving a most alarming and unusual bark. She wasn't very big, but as fast as lightning. I jumped out of the pool, to give her time to get used to her new surroundings. I then gave her a few fish, surprisingly, she took them gently out of my hand. I moved the empty crate out, and opened Kim's door. He was much more frightening than Judy and flew out giving a loud, deep bark. At this I rushed out of the pool, but, to my amazement, so did he! I didn't realize sea-lions were so agile. Judy didn't find it quite so easy, but after a little straining, she too was out. I didn't know which way to turn, if they had been grey seals I would not have been frightened, but these beasts could move faster than I could.

Kim climbed up my rockery and before I knew what he was at, he was on top of my garage, which, at the back, was level with the car park. Another second and he was running around the cars. I grabbed his cage and ran after him. Finally some visitors helped me to drive him into his cage. This meant I must at once put up netting to keep the two sea-lions in safety.

It took two days to make the whole Sanctuary secure; meanwhile I set up boards high enough to stop them from climbing out.

I filled the pool and watched the sea-lions' graceful action in water and admired the ease with which they flew out of the pool. When they were ashore they slid along the smooth slabs; they enjoyed this so much that they soon started taking short runs, falling on their tummy and sliding along the side of the pool from one end to the other.

I had an old mirror, which I fixed on to the wall and they would slide up to it, fighting for a good position from which to admire themselves, after which they would walk backwards into the pool. I suppose they were fascinated by seeing another sea-lion in the mirror backing away from them.

Scratching was another of their entertainments; they would sit on their bottoms, lift one back flipper and scratch their chins. They were too nervous to allow me to stroke them, but they never attempted to bite me, though they would come roaring past and knock me over if I were not careful.

I made a strong wooden platform for them to sleep on, but no, they preferred to sleep in their cages which I had placed inside the shed.

Judy and Kim were given treatment once a week because they were covered with spots, these were caused by the migration of a worm around their bodies. They also coughed and brought up worms from their lungs. One day I found a tape worm three feet long in the pool. When I first saw it I thought that some-one had thrown a shoelace in.

Judy seemed to be the boss, she was very mischievous; Kim liked his afternoon nap in the shed, but Judy would soon turn him out in spite of his grunts of dis-pleasure at having his sleep disturbed.

The barks of the two sea-lions were worse than their bites. They often made a lot of noise even when playing, and looked as though they were fighting, but this was their way of enjoying themselves.

The main thing now was to keep down the worms for if they penetrated the lungs deeply, then the sea-lions would certainly die.

CHAPTER NINE

Simon II and Sally II

It was again September and, if the weather was rough, we could expect some pups to come in soon.

The first came on the 2nd of October, it was a pup about a week old. As the business was closed we had all the time we needed to care for it.

We gave it milk feeds four times a day for about a week, by which time a second pup had arrived.

Often we imagine that we are going to have an easy winter, but in fact, we usually end up with roughly the same number of pups; the ages, of course, vary from a day old to two or three weeks but nearly all of them have not yet been weaned.

Baby seals, while they still have their creamy white coats, look alike; it is when they lose these that one is able to tell them apart even from a distance. Each one then has its own characteristic appearance. But even after this, local people still ask which is which, and to them they look the same, though when you have been caring for them, you feel like a person with ten children, who, even if they look alike to strangers, present no problem of identification to their parents.

These two pups were successfully weaned on to fish and returned to sea, as were two others that came in a little later, but one that arrived during November stayed with us. We called him Simon II, because he had the same temperament, and looked very much like our old Simon; he even had the same complaint, pneumonia, which was going to need long-term treatment. We gave him antibiotics, and kept him indoors for quite a long period. Weaning him on to fish proved no problem.

It was January when he was first introduced to the pool and also to a meeting with Judy and Kim. I didn't know what would happen when we put a grey seal in with the sea-lions, so I was prepared for any emergency.

Almost immediately Simon II and Judy's noses came together, then he made a fast getaway, but Judy did two circles around him before he reached the end of the pool. Kim took no notice of him.

Judy's speed was terrific, so at feeding times I had to throw her fish up to the far end of the pool, then quickly throw Simon's just beside me, or she would get the lot. Kim usually took his out of my hand.

In March of that year, a woman from a nearby resort drove over to tell us that there was a badly-injured seal on their beach. I followed her back in my car; she had about six children with her, and they were all much excited.

I expected the seal to be either an adult or about five to six months old, as I had never found a young pup at this time of year. As the tide was out, we had a long walk over the beach; when we finally got to the place where the seal had been, I could not see a sign of it so I walked up to the base of the cliff. There, in between some rocks, I saw a small pup, about three weeks old. It was thin and, looking closely at it, I noticed that its neck was torn open, either someone had put a rope around it, or it had been caught in a fishing line – it looked really bad. As I went to pick it up it started to move away so I let it stop of its own accord, then I put one arm under its chin and the other under its tummy. I tried to be as careful as possible as I didn't wish to cause any further damage to its neck.

I saw that it was a female, and was suffering from shock. I had a long way to carry her along the beach, and I thought 'this is where a jeep ambulance would be useful', for with it treatment could have been given on the spot and warmth and comfort provided at once.

I had to rest at intervals, putting the seal down on the sands while I did so. The children kept asking if she would die; I assured them that she would have the best possible care and that I thought she would get well.

When we reached the car, I put her on the seat, thanked the children and the woman who had come with us and told them they could visit the Sanctuary at any time to see her. Then I got home as quickly as I could, made a bed of straw, switched on the infra-red lamp and put it in a safe position where the seal couldn't get at it. I now had a good look at the pup's neck, and then rang my vet and gave him all the details of the case. A little later he arrived, checked her over and gave her an injection.

That night we had no sleep as the pup's breathing had worsened and it looked as if we might lose her. We didn't put her on to milk, but weaned her straight on to fish: small quantities every two or three hours.

Next day, the vet came again and gave the pup a thorough examination which showed that she was in a very bad condition. Her lungs were severely congested, indeed, she was now gasping through her mouth for air. Her two gashes were deep and infected, they were close together and had jagged edges, so the vet said he must treat the injuries as an open wound for they were too wide to stitch up. The cuts were thoroughly cleaned and sprayed with antiseptic and every few hours I dabbed gently at them with antiseptic powder and cotton wool, getting the powder well down into the wound. The pup was very good while I treated her neck, as if she knew that I was trying to help her. Many more sleepless nights followed that first one.

At one time her breathing was so bad, and her nose so blocked up that we decided to use Vick to loosen and bring up the mucus. I placed it regularly on her nose and it proved successful. Although she had let me treat the wound, I had to use a stick to put the Vick on and she snapped when I tried to stroke her tummy.

Sally II, as we decided to call her, was in for a long spell of intensive care. Gradually the wound dried up and began to heal; her breathing remained bad, but the Vick made it possible for her to breathe at times through her nose.

Sally must have been just over three weeks old when I picked her up for she had recently lost her puppy coat.

Weaning her on to fish was awkward, as we didn't want to hurt her while holding her in order to force the fish down. Normally, pups have to be held quite firmly while doing this, but in Sally's case we had to be very gentle. However, each day feeding became a little easier and we thought her condition must have improved as she snapped out after putting down her food. This is a normal thing for a healthy pup to do; affection is not gained quickly, trust has to be built up on each side and this takes time.

After five weeks I moved Sally to another pen; her breathing was slightly easier but she was still passing mucus from her nose. When she heard the other seals splashing in the pool, she used to climb the netting and obviously wanted to see what was going on. On the way to their sleeping pen Judy and Kim had to pass the netting and that was how she first met them. I felt that Sally II might be going to be yet another Simon. For one thing, I knew she would never be able to be returned to the sea, her lungs were sure to remain weak just as his had.

After hardening her off for a few weeks, I introduced her to the pool. She slithered into the water, then there was a commotion, flippers were flying everywhere. She jumped out at my feet and let me stroke her, after which she went back into the water. Simon II chased her round the pool, sniffing at her back as they moved around; Judy also joined in, but not Kim, he went back to bed, as if to say 'not another!'

I kept close watch on Sally, in case one of the seals turned on her, but in fact she was accepted as if she belonged among them. Simon was certainly excited and I began to think it meant love at first sight; Judy, perhaps disapproving, just kept pecking at her. I smiled to myself as I watched Simon chasing Sally around the pool and she, popping her head up occasionally to look at me with her big bright eyes, plainly wondering how she could throw off this big fat hulk of a seal. I was very pleased to see her playing as though she had never been ill. This was one of the moments that made me feel on top of the world, and convinced me that the job of rescuing seals can be rewarding as well as, at times, disheartening.

CHAPTER TEN

The suffering that people can cause

Now it was time to get my business going, for Whitsun starts the season and many holidaymakers arrive from then until September. It was during the next few weeks that the vandals started up again. First, some ornaments were stolen, then things were thrown into the pool; children were allowed to paddle in the fish pond and my outside wall was kicked down by people climbing over it.

By the time August came, I was preparing to stop the public from entering the Sanctuary and for one brief moment I was nearly ready to stop having one. Sally was hit on the head by a brick, one of several which had been thrown from the car park behind the canopy into the pool, and Simon had swallowed a ball, which had been thrown into the pool by a child whose parents were present at the time. (I didn't see this myself, but a visitor came straight down to the shop to inform me.)

Why do people do these things? I had signs telling them not to throw anything into the pool and also telling them of previous damage caused by such actions.

I rushed up to the pool and first looked at Sally; she was holding her head above water, her eyes bulging. Normally, as I passed the netting surrounding the pool, she would jump out, but now she seemed to be staring into space. There was a mark on the top of her head, but no blood. The first thing I suspected was concussion. Next I looked at Simon; he seemed all right, and I supposed that to him swallowing a ball was like swallowing a fish. I checked with the visitors that they had actually seen him swallow the ball, they all confirmed that he had.

I emptied the pool so that I could check Sally over more thoroughly; there was no cut, but there was a depression in her skull. I pressed lightly on it to see if this caused her any pain; it was difficult to judge whether it did, more still to tell what damage had been done. I went back to the shop, telephoned to my vet and told him that neither seal seemed to be in any pain, but that Sally probably had concussion and that Simon had swallowed a ball. The exact size of the ball was not known, my informants said it was a little larger than a golf ball. What I was worried about now was whether the ball would block his stomach as the straw had done to Sammy. We discussed what could be done to help Sally; if brain damage had occurred then we were in serious trouble, for an operation would have meant certain death.

We decided first to dose Simon with liquid paraffin; this would give lubrication and perhaps help the ball to pass along the intestines. So far as Sally was concerned, we must wait to see if any further symptoms developed, if they did they could help us to diagnose her injury; she might only be stunned and, if so, would recover quickly.

I went back to the pool and fussed Sally, but she didn't seem to know I was there. At that moment, if I could have laid my hands on the person responsible for her state, I'm afraid I would have taken the law into my own hands and who knows what the outcome would have been. A friendly neighbour volunteered to keep watch on both the seals while I went back to the shop, but every half-hour I popped out to see how they were. I was anxious about the ball; if Simon were going to pass it how long would that take? Luckily he was still eating so I was able to put liquid paraffin into his fish and, if he didn't eat, he was so fat that if the ball blocked his stomach, he could live quite a while on his own fat. Sally ate very little, but I wasn't too worried about that; she too had had a good appetite and was well able to starve for a few days without its having any ill effects.

Before I went to bed, I spent some time watching them. I might as well not have gone to bed, for Mary and I stayed awake and I was up early checking again. Simon looked up as I entered the pool, which was still empty, making it easy for me to check the motion he passed and also rest Sally without having to distress her by carrying her into isolation. In a case such as this one, company would do wonders especially as Sally was probably more sick in mind than in body.

She lay motionless; suspecting the worst I jumped in quickly, afraid to touch her in case she were dead. I stroked her gently, her body was warm; then she lifted her head and laid it down again as if she hadn't seen me, no doubt she didn't see me. I spoke to her; Sally's ears remained closed; I wondered whether she could hear me; I didn't think she could. She seemed to be blind and deaf too.

I felt sick. What had they done to the poor little thing? As if she hadn't gone through enough already. Tears came to my eyes at the thought of the stupidity of the people who could hurt an animal in such a way, and I realized, with a shock, that this could have happened to my dog or even to my daughter if irresponsible people threw bricks. Now it was up to me to make the Sanctuary safe from any vandals without spoiling the pleasures of the many people who were taking an interest in the seals. I had received letters from a school in Holland whose children had sent a small donation collected by selling old toys and sweets (their teacher comes over each year to take photographs). Canada was another part of the world from which I had had letters, many had come at the time of the culling of the pups in the Gulf of St Lawrence. From Malawi, too, I had had encouraging support.

That day brought no change; Simon didn't seem to suffer any ill effects, but Sally showed no interest in anything that went on. I posted a notice explaining what had happened to the seals.

Forty-eight hours went by and Simon still hadn't passed the ball. The following afternoon, after seventy hours, I was relieved to see it appear; it was solid rubber with ragged edges and was the size of a golf ball. Had it been any bigger I don't think Simon would have lived.

Now we had Sally to hope for. I thought I'd like to see if she had any reaction if the pool were filled, so I filled it that night to Simon's great delight. Kim and Judy too, were pleased.

Simon at once started teasing Sally, but he got no response; she sat up in the pool staring into space, then she suddenly gave a yell and spun round like a top. This amused the spectators, who thought it was an act; when I told them what had happened to her their reaction was different. I shouted to Sally but she did not respond. When Simon again went near her she flipped him and gave a sudden dive to the other end of the pool. She now ate a little but only if I put the fish into her mouth; if it dropped out she left it, she didn't seem to see it. However, she was at least getting sufficient food to keep her going. There was nothing we could do for her surgically, so love and affection, and the passing of time, would offer her only hope of a cure.

Judy, Kim and Simon did their best to entertain her, but nothing seemed to have any effect on her. Friends kept watch whilst I was busy; they told me how many times she did her spinning act. At feeding time she took a little fish, but did not show much interest in it. I kept on telling Simon to show her some affection, and he must have sensed what I meant, as he slowly approached her, gently sniffing her neck and then lay on his back, brushing up against her. This state of affairs went on for a few weeks, after this Sally gradually came to the side of the pool and resting on her back, let me stroke her stomach. She now ate much better and started moving around the pool a bit. One evening when I closed the shop, I went in to inspect her, and found that Simon was cuddling her, and that she seemed to be enjoying herself. I crept quietly away. This, I thought, could be the turning point.

Next morning she ate like her old self, even fighting Simon for fish. I was very pleased. She had been through a rough time from her birth onwards, and I was determined that no further harm should come to her. I set up a canopy over part of the pool which would prevent stones from being thrown in and also provide a shelter for the seals; over the other part I placed netting. The seals were now safe, as well as happy.

CHAPTER ELEVEN

A disastrous winter

SEPTEMBER 1969 started off in the usual way. When a pup was washed in on our beach a friend woke us up at seven o'clock in the morning, he was carrying a little seal in his arms. It was about two or three days old and remains of the umbilical cord could still be seen; its bright eyes looked up at me, while it rested its head on my friend's forearm. The pup had patches of oil on its creamy white fur; it weighed about twenty-five to thirty pounds which meant it was in good condition as regards weight, but it was beginning to run at the nose showing that congestion of the lungs had developed, which if not treated would lead to pneumonia.

The shed was ready, all I had to do was put down some straw; this took only a few minutes and then the pup was placed on to a warm bed. I fixed up the infra-red lamp and started treatment immediately. The next thing was to inject it with an antibiotic; I rang my vet who came at once. We injected into the fatty part near the hind quarters and further doses were given on the following day in its milk feed. All that was now needed was warmth, regular food and rest.

Four days later a couple from St Ives rang us up to say that they had just visited Hell's Mouth and seen a baby seal in difficulty on the beach. A young fisherman was keeping his eye on it while they were phoning me. The fisherman had told them that the pup was in a terrible state; it looked as though its stomach were hanging out. We asked the couple if, so as to avoid any delay in our finding it, they could stay at the place where we would have to turn off the road.

We put our usual gear into the car and raced away. The twelve miles we had to go didn't take us long. We took the car down the lane as far as we could, then started off on foot.

We had a long walk before reaching the top of the cliff; then I saw a narrow path with only just enough room for us to walk down it in single file. It was a very steep slope indeed with a dangerous drop on the seaward side and we had to be very cautious.

I had a jacket in which I intended to wrap the seal; I put this on my outside arm, using the other to hold on to the grass in case I slipped.

Mary was with me when I started from the car and so were the young couple who had rung us up. I looked back up the slope and saw that they had just reached the cliff's edge. On looking down again I saw the young fisherman by the seal and I noted that the tide was only a few feet away, so I made as much haste as I could with safety.

When I reached the boy, he asked me if I were going to put the pup out of its misery. I examined it; the sea had just reached where it lay so I gently turned it over. I was expecting a frightful sight but all I saw was the umbilical cord hanging out; the lad had evidently mistaken this for the seal's stomach. While I was making further checks a wave came rolling in and drenched me, it also started to drag the pup away on its backwash. I had to be quick now or it would be taken out to sea. I picked it up just in time as another roller came in. In fact, I caught the full force of the wave and only managed with some difficulty to hold my feet firm.

While all this was going on, Mary and the others had reached the beach; they feared I was going to be washed out to sea when that last wave hit me.

I looked up the path we had to climb and knew I was in for a rough time. Carrying thirty pounds of seal up a three-in-one slope, where there was only just room for foothold and a drop of a few hundred feet down to the sea, was going to be a little breathtaking. I made the baby as comfortable as I could across my arms and started on the long pull up. Each step I took the seal seemed to get heavier; sometimes its back flippers would touch the gorse on the bank and then it would wriggle to get free. This made things worse for me, as my arms were getting tired, so I got down on one knee holding tight on to the pup. The cliff looked frightening and I knew that the least movement might send us both tumbling down. My breathing was very difficult and I had to rest every few minutes before I started off on the climb again.

I was happy when I reached the flat top. There was still quite a way to go, but the worst was over, though by now I was completely breathless and when I was well on the grass, I put the seal down.

Mary said I looked like a ghost and I certainly felt like one; but after a few minutes I was able to make for the car. We put the pup on the back seat. It looked healthy enough so we hoped we would be able to prepare it for the sea. On arriving home, I made a clean bed of straw and settled the baby down on it. Just in case of any infections I called the vet out to inject it and then it was left to rest. Two hours later the pup was dead. We couldn't understand it.

A post-mortem was carried out and showed congestion of the lungs and septicaemia. Septicaemia is a form of blood poisoning and was secondary to its other condition; it was the first time I had come across it in seals, but later it became very common and nearly seventy pups were to die of starvation, lung congestion and septicaemia in the months that followed.

My next call was from a man living thirty miles up the coastline; he had found a baby seal only a few days old and wanted to know how to feed it. He wished to do this himself so I explained what he would need to do.

The following day I rang him up to see how he had managed as I knew it wouldn't be easy for him. He told me the pup had died four hours after he had found it. Again the cause of death proved to be septicaemia and again this one had looked a healthy pup.

Further calls came from St Ives stating that three seals had been washed in dead. Then came one from the Lizard saying that a baby seal had been washed in with its eyes destroyed. I arranged to meet the man who had found it, half way to the Lizard so that I could get to my vet more quickly. They had put the seal in a net and it was easy to handle; I had a quick look, the eyes looked very bad, and the seal was very thin. It had already lost its white coat but I do not think it weighed more than fifteen pounds; at this age it should have been around sixty to eighty pounds.

I went straight to the vet as the pup's condition was so serious and it was also suffering from shock. We laid it gently on the table, and while doing so one eye-ball ran out on to the table so that eye was completely gone. The other wasn't quite so bad and the vet injected it against all likely infections. I rushed it back to the Sanctuary, but the following day it was dead; the post-mortem again showed malnutrition, lung congestion and septicaemia. What was happening? We were unable to save these pups. Why were this year's pups different from earlier ones? I began to worry and after a few more phone calls telling me of other baby seals coming ashore at the Lizard with what looked rather like burns and of still others gasping for breath, I decided to call in the Seal Research Unit to see if they had any views as to what might be causing these deaths.

Soon after this, television and the press carried the story. Our telephone began to ring non-stop, our house became something like a newspaper office and we were surrounded by television cameras. Since we were at the time full up with starving pups who needed the most intensive care and we were also being constantly called out to rescue others who had been washed in, we had quite a problem to get through all our tasks and give interviews as well.

Mr Bonner, of the Seal Research Unit, and I decided that we ought to make a survey of as many beaches as we could.

Two journalists from the *Sun* and two from the *Evening Standard* volunteered to help us.

We started on our local beaches and worked our way down the coastline by Chapel Porth, Porthtowan, Portreath and Godrevy from which we had had a report that there were two baby seals on the beach. We contacted the man who had informed us and he gave us directions which we followed.

We had to go by a place where after a steep slope there was a terrific sheer drop, it was practically vertical and to look down was frightening.

We eased our way along the first part but the final drop was going to be very difficult. I had a short length of rope with me and was lowered down as far as it reached, after that, I made my way as best I could down the last lap.

I could see one baby seal lying between two rocks, the other was in the water, both looked very thin. I could do nothing about the one that was in the sea and just hoped that it would wean itself on to fish before it was too late.

The pup between the rocks was in a very bad condition and had lung congestion. I placed my old jacket round it and made a comfortable hold round its body to enable me to clamber back over the rocks to the base of the cliff. I saw the reporters half way down the rockface waiting for me, then one was lowered so that he could take photographs of the rescue.

My problem was how to get the seal up the sheer part of the cliff. The rope would be of no use and I would need help. I carried the pup as far as I could. Then I asked the photographer to stop taking shots and to hold the seal. First, I wrapped my jacket round its head so that the man should not get bitten, then I explained to him what he had to do and handed him the pup.

I had quite a struggle to reach the first ledge. When I had made it, I dug a good foothold in the shale. Then I reached down to take the pup from the photographer. I stretched as far as I safely could, an inch more and I should have toppled over, but I could not reach the seal. I asked the photographer to try and lift it as high as he could. By now I had been joined by a reporter, he dug in his foothold behind me and held on to the back of my coat to prevent me from slipping. Like this I was able to stretch a little farther and managed to get one hand on to the seal's back flipper. I held on like grim death, for I knew it was now or never. The reporter hung on to my jacket as hard as he could, the photographer pushed up as high as he could but then he shouted that he was afraid he'd have to let go. The pup was now hanging by its flipper which I was holding, and its weight was getting too much for me.

I called to the photographer to try to put the jacket round its head to prevent it from twisting and snarling. Then I heard a yell, he had been bitten on the thumb. Nevertheless he managed to cover the pup's head and to ease the weight and I heaved it on to the ledge. After this I crawled up the next hundred feet.

Even when we had reached the top of the cliff our troubles were not over for the car was three-quarters of a mile away and for most of that distance we had to walk through thick brambles, this made it impossible for me to put the pup down and give myself a rest until we reached a field. Over the last few hundred yards I had to rest many times but I made it.

I had expected the reporters to catch up with me but there was no sign of them and I went back to the top of the cliff to have a look. There I discovered that the photographer was stuck on a ledge, so I crawled down to him, after which we pulled him up slowly.

Finally we all went back to the Sanctuary.

The newspaper men had risked their lives to try to save a seal pup, and one would have a scarred thumb to remind him of his good deed; they also had a big story and photographs to prove its truth. It is sad that the pup, like most of the others we had rescued during this season, died of lung congestion, septicaemia and starvation.

Seals still went on coming in on most beaches, some dead, others in a very bad state; one came from St Austell. We had never before had reports of seals washed in there; again this one was dying of starvation.

We rushed it back for treatment. The vet had little hope for it, but we hoped something could be done. Already nearly thirty pups had died and this was four times the normal number of pups washed in at this season.

Scientists came down to investigate; checks were made on seals washed in with burn marks, also on fish that had ulcers. Every possible cause was considered.

Many thousands of sea birds had died in a similar way to these baby seals; the post-mortems and tests on the birds had shown that PCB and DDT were present in high quantities. Quite a number of my seal pups died in convulsions. It was pitiful to see them; they had used up all their fat until they were only skin and bone, and no doubt this had left them open to any disease. Were the seals being poisoned by effluent from some plant? Were the fish being poisoned?

It was easy to blame every sort of pollution. Had the detergents to clean the beaches of the *Torrey Canyon* oil been involved in what had happened and was still happening?

We hoped that the scientists would find the answer; in the meantime our object was to treat the sick seals, but how could we do this without knowing the cause of their condition? The scientists worked hard, but in the end they failed to find any evidence of pollution being involved.

From having seven to ten pups to care for each winter, we were now over the forty mark – and how many more were going to come?

All Cornwall was as worried as we were, for one of our main industries is fishing and we didn't want people panicking about poisoned fish.

We checked on the reports of seals with burns, but none were found. Then we decided to hire a boat and again investigate the coastline and the sea close to the cliffs to see how many more babies might be floating around in trouble.

The boat pulled out of harbour at nine thirty in the morning. We had hoped for a good day, but the north-westerly wind blew and we began to rock. Like a fool, I went below where there was a new oil stove belching fumes. The sea tossed us about like a cardboard box and I could hardly hold on to the seat. I soon started to sweat and felt sick, so I went quickly on deck where, after a few whiffs of fresh air, I was all right again.

The pathologist of the Seal Research Unit was with us; we made our first stop at Seal Island. In the summer there are always quite a number of seals basking in the sun on the rocks, and many visitors take trips out to the Island to see them; but now there wasn't a seal in sight. We had a good look and then went farther along the coastline.

We used powerful glasses to check the caves and beaches normally inaccessible to man. We couldn't see any bodies, but suddenly we saw a little head popping up out of the water. That seal at least seemed to be all right – indeed, he was very inquisitive and kept watch on us.

I thought, 'At least one has survived, even though we have lost nearly a whole generation of pups.' We carried on surveying the cliff line and the water's edge; in all we saw five seals, all young ones, but not one adult was in sight.

We decided to call it a day, the skipper turned the boat around, and we headed home. When we arrived near the harbour we had to wait an hour for the tide to give us sufficient water to make it. We were cold and hungry, but we had satisfied ourselves that no adult seals were in trouble.

I was anxious to get home as we had four pups at the Sanctuary. We always name our seals; the new arrivals were called Flipper, Jenny, Fatima and Blacky.

They were on fish, but Fatima and Blacky still had to be force fed; it was a full-time job. The fish were prepared, Sally, Simon, Judy and Kim were fed. Then I concentrated on the babies. Flipper and Jenny were fed next; they took the fish from my hand, but were kept in their pens for the moment as they had a long way to go before they would be back to normal.

Fatima's meal came third, and now Mary had to help. I got into the pen, held the seal's head high, forced her mouth open and Mary pushed the fish down. At this stage, the pup was having seven mackerel three times a day; this made up between five and six pounds of fish. Blacky was put in a separate pen, away from the others; he was still a very sick seal with a very bad eye. He had a queer habit of charging at me snarling and then suddenly rolling over. It was undoubtedly a game as he did it a few times before and after each feed, but I couldn't take any chances with him for very often he would tear at my coat – as long as it wasn't my skin I didn't mind.

Blacky was given the same amount of fish as Fatima; this was a little less than Flipper and Jenny, yet Blacky and Fatima seemed to put on more fat than the two others. Probably this was because they had had a better start in life than the little ones. It wasn't easy to feed Jenny and Flipper as, although I had fed Judy and Sally before them, they would jump out of the pool near the babies' pens and try to take the fish out of my hand; and this would raise a crying snarl from Flipper and Jenny.

After a week I decided to let Flipper into the pool; I opened the door and out he came following in my footsteps. He sniffed at the water, but that was all, he was more interested in getting a little fuss from me. Finally, I had to give him a push, and splash – he had his first swim. With his fur wet, he looked tiny beside the others. They swam up to have a look at the new arrival and all were gentle with him, but he was having none of it; he splashed the water with his front flippers, gave his cry, and tried to jump out of the pool. Although the water was only six inches from the top, he found this very difficult and it took him a few hours before he finally made it. After that he soon learnt how to hop out.

The following day I lowered Jenny into the pool. She slithered straight in near Flipper, splashed everywhere and the two of them had a good old scrap, but they did not hurt one another. The next problem was how to feed them while in the water. They would eat out of my hand, but couldn't yet eat fish thrown into the pool. I had to throw the other seals' fish to the top end, then quickly hand a fish each down to Flipper and Jenny; as I did so they jumped up at it and my fingers nearly went with the fish. Very often people looking at the Sanctuary would ask me questions whilst I was feeding the pups and I had to take care not to have my attention distracted.

The doors to Flipper's and Jenny's pens were left open and, at night, they used to go in on their own. One evening whilst we were in bed, we heard a great clatter in the Sanctuary, so I rushed out to see what was going on. I found that Judy had broken down the door to Fatima's pen and she was in the pool. This was going to be very awkward as she hadn't yet learned to take fish out of my hand.

Next morning I discovered her trying to take down a fish that one of the others had left; finally she succeeded. She had done what Flipper and Jenny couldn't yet do: pick up a fish from the water.

There were now seven seals in the pool and one in the shed. The main thing was to see that each had its fair share and this took some sorting out. Flipper and Kim nearly always jumped out for theirs. Fatima sprawled against the gate about nine feet from where I usually fed the others, so I could see to it that she had her share. That left Judy, Simon, Sally and Jenny in the water. I'd throw a fish to the top end of the pool for Judy, then quickly throw one to Sally and Simon, leaving Jenny close to me giving her first cry which enabled me to drop a fish into her mouth. While all this was going on, Kim and Flipper were around my legs pulling at my trousers and wellington boots; then Flipper (his cry was terrific, and I think it could be heard at the top of the valley) would keep at me until he had his fair share.

Finally I would go to the pen and feed Blacky who just wouldn't eat on his own. I had once put him into the pool, but he had no idea of taking fish in the water and with all the other seals swimming around, if I threw the fish to him, someone else had it before he could even turn to smell it. When, after two days, he still didn't attempt to go for the fish, I had to put him back into the shed as I couldn't leave him long without food.

My time was pretty well occupied. After feeding the seals I got the fish out of the fridge ready for the next feed and then started looking after our birds. We had twelve small parrots whose cages needed to be cleaned out and fresh water and food put in, and then there were the two macaws (Blue and Gold) who needed the same attention. Harry and Mac, as they were called, usually said 'Hello' to me, after which I scratched their necks and Mac always wanted to come out on to my shoulder, to peck at my ear and pull my hair.

While I was dealing with the birds, out came the badgers from their home-made set. They had been found abandoned while still too young to feed themselves. No doubt their parents had been killed by people who regard badgers as enemies. They were brought to me to be weaned and cared for and then settled in for good. First they sniffed around me. (They recognized my scent; had I been a stranger, their hair would have bristled out like a hedgehog and their heads would have been tucked down low.) Then they went to the tins into which I had scraped the droppings and the food I had cleaned out of the birds' cages. These they pulled over with their thick, strong front legs, and then scratched out all the seed.

During the day the badgers accepted my company, but were not very playful. It was not till ten o'clock in the evening that, even though I had the lights on, they would begin tearing around, pulling at my trousers and jacket and charging to and fro between my legs. This was a regular game that took place each evening and to which they looked forward.

Once they had all been fed, I had to see to the two mynah birds who lived in our house (by now they would both be chattering noisily). They could say almost anything, talking as a rule in the tone of the person who has taught them their words.

One way and another, we now had a large and varied animal family.

CHAPTER TWELVE

Protection for the wild-life of our coasts

IN March 1970 a few sea birds – mostly guillemots –
were brought to me with oil on their feathers; not badly
oiled, but sufficient to make them waterlogged, so they
were either carried in by the current or they made for
rocks near the water's edge.

Our beach at St Agnes quickly became a death-bed.
There were thirty dead guillemots lying thickly covered
with freshly deposited oil, also two gannets and two
razorbills. Gannets are large sea birds with long beaks,
very similar to young swans. On the rocks I could see
about another thirty birds huddling together in little
black groups. These proved to be guillemots, the black
oil had obliterated the whiteness of their feathers and
the grey of their head and wings. The stance of these
birds reminds one of penguins and they looked like a
group of sad little old men.

I knew it was impossible for me to cope with such
a number in my Sanctuary, so I rang the RSPCA
Inspector for help and to ask him to make arrangements for them to be taken to the Bird Sanctuary at
Mousehole, near Penzance, where the best treatment
would be given to them.

The Inspector came immediately and that morning
we gathered forty guillemots from the beach and the
rocks. It was pitiful to see them: the majority were
completely plastered and so thick was the oil on their
bodies, they could not even run away from us. We
got them back to the Sanctuary, transferred them into
special boxes and made arrangements for their journey.

Lunch over, we made a trip to Trevellas, another little

beach off St Agnes. There we found another ten birds dead and six alive, but badly oiled. Again we found two gannets, this was unusual for during the *Torrey Canyon* disaster when thousands of sea birds lost their lives, only two or three gannets were involved.

Soon birds began to arrive from Perranporth, a few miles away, and then from Newquay. People on the beaches noticed their plight and began collecting them and bringing them to the Seal Sanctuary. During the afternoon, we collected another fifty birds off our own beach, all in a very bad condition. This was serious, but to any outsider watching us, it must have appeared comical; there was the Inspector and myself chasing the birds, who went first one way and then the other, always trying to make for the safety of the open sea. It was necessary for us to catch them quickly before they started preening, for if they did this feather-cleaning operation themselves, they would swallow oil which would burn their stomachs and death would be certain. Another hazard was provided by the gulls. The oiled birds were a helpless target and were often attacked and pecked to death by the tougher seagulls.

For the second time in three months my garage was housing distressed sea victims; first the dying baby seals and now the dying sea birds. The ones we knew had no chance of survival, and who were suffering, were put to sleep. By evening, our total of dead and alive birds rose to 163, including eight gannets. We made a final check on the beach as darkness fell : there were still many birds out at sea sitting on the water and we knew that tomorrow would again be a busy day.

At nine o'clock next morning I saw at least twenty birds on the sands; the rocks to the left were obscured from my view, but I guessed there would be as many again sheltering there. By midday we had picked up eighty oiled victims. We had covered the three beaches in the St Agnes area, but it was easy to miss birds, especially if they were among the rocks, for the oil acted as camouflage, birds and rocks blending in colour, and also guillemots were apt to wedge themselves in crevices and only the occasional flutter gave away their hiding place.

The beaches were not one mass of oil, as with the *Torrey Canyon*, but small amounts were being deposited by the tide as though they had been separated from a larger slick off the coastline. The winds were strong north-westerly, helpfully washing the birds in. Other areas were sending to us for help, showing that the affected coastline stretched from Newquay to the north of us to St Ives to the south.

The count that evening showed that still another 150 birds had been collected and these included four gannets. By then we were feeling the strain, for one thing our legs were unaccustomed to these miles of walking and sprinting.

Next day we found that overnight more oiled birds had beached themselves. They needed to be picked up. At this point, the Inspector and I realized that to cope with the beaches and to reach all the distressed birds was getting beyond us. An appeal for volunteers to patrol the beaches and to help transport the birds to Mousehole was made on television. Many answered the call.

At present, by English law there is nothing to stop ships depositing oil into the sea outside a limit of three miles off the coast. The captain of a vessel intending to do this may, if he wishes, inform the nearest coastguard station that he is depositing oil from his bilges. What follows is then up to them. In this way the coastline can be polluted, thousands of birds can lose their lives and the ratepayer, through his Council, has to foot the cost of cleaning the beaches. This is surely a most unsatisfactory situation.

People in Cornwall were by now taking me for granted and any animal or bird injured, or in difficulty, was brought to me – I was anxious to give all the help I could, but I hadn't enough room to take care of all the casualties properly. So to have the funds to extend the Sanctuary, we thought that the only solution would be to sell our beach business.

During the winter of 1970 all the pups washed in developed ulcers, in some cases the bodies were covered all over with them. Specimens were taken but tests revealed no specific cause. Mrs Joyce Butler, the MP for Wood Green, asked a question in the House of Commons about the recent deaths of seals – nearly a whole generation of baby seals had been wiped out. But the Minister of Agriculture and Fisheries was only able to reply that the investigation he had ordered to be undertaken by the Veterinary Investigation Laboratory at Truro had found no evidence of contamination by any toxic substance and that the findings had been passed on to the Natural Environment Research Council whose Seal Unit had the matter in hand.

When the scientists gave their verdict it was only to state, as before, that the seals had died of starvation, lung congestion and septicaemia; they could offer no reason why so many had been so badly affected in relatively mild climatic conditions. But whether the cause was some unexplained natural disaster or due to man's careless pollution of the sea I was determined to be ready if it should happen again. I would sell my beach café and bungalow to raise the money to build a properly equipped Seal Sanctuary.

I took the decision which was eventually to lead to the confrontation in Gweek fully conscious that I would be mortgaging my family's resources for years ahead and reasonably prepared for the difficulties which

I might have to overcome. It was something I had to do and events, or perhaps fate, had been pushing me in that direction for a long time. Ever since the rescue of our first 'orphan' it had become taken for granted, all over Cornwall, that any animal or bird injured, or in difficulty, was brought to me. In this way I had even acquired some badgers who had been found abandoned while still too young to feed themselves. I had responded to all the calls for help but it was obvious that the pools I had built on my lawns were reaching the limit of their capacity. I now had seven seals (who were growing larger each year) in the pools and one in the shed. I could not put permanently sick and injured animals back into the sea to face death a second time. And I knew that every winter would bring more 'orphans' to care for. Over the years I had learned a great deal about treating the diseases seals were subject to but with better facilities more could be saved. And more babies could be weaned, taught to eat fish and returned to the sea in a better position to fend for themselves. I had worked out what was needed, a site large enough to accommodate at least three good-sized pools, a filtration plant which would keep the water clear and free from bacteria; outbuildings to accommodate sick pups with infra-red lamps in each room, a hospital to treat the injured and a deep freeze to store fish. Whatever the cost nothing was going to stop me putting my plan into action. Not even the last-ditch opposition at Gweek.

CHAPTER THIRTEEN

Completing the new Sanctuary

THE television programme described in the first chapter brought a great groundswell of support from all over the West Country but it was another hard-fought two years before my original plans were once more agreed and I was given permission to recommence work. Even at the last minute the whole scheme nearly foundered over a licence to build a public toilet. Without this amenity the Tourist Board would not grant my loan but the Planners refused all the sites I proposed. Impasse had been reached when I had a flash of inspiration. I asked to be allowed to extend our existing staff toilet and they could find no reason to refuse this. So, by a ruse, we got our toilet facilities for the public. It was now October, 1974 and we had only seven months to complete all the work if we were going to be ready to open to the public at Whitsun, 1975. It was going to be a battle not only against time but against inflation; costs of materials, labour and building had risen in a frightening way over the last two years. A lot of work would have to be done by myself and any friends I could muster.

I was lucky that a good friend of mine from Perranporth immediately offered his help on a low rate of pay; we arranged to start work Monday morning the first week of October. As I was still operating from St Agnes I made the 24-mile journey every day, getting up at seven to feed the seals and prepare more fish ready for when I came back at 4.30 p.m. Mary gave the babies a midday feed as they needed more regular feeding.

A local friend at Gweek passed on any urgent phone calls, especially those reporting baby seals washed in. My mate, Jack Knowles, made his own way each day from Perranporth. At nine thirty the lorries started to arrive with the granite lumps, these were dumped in piles at the end of the pools. A digger was on site, so we used this to push the large granite pieces as near to the pools as possible, then came the hard work of using bars to manoeuvre them into good positions. Facing the sides of the caves proved to be the most difficult job as we wanted the entrances to blend in with the opening to the concrete interior. We used the digger as much as we could then used steel bars to place them properly. One lump of granite weighing a ton and a half would have crushed us if we made the wrong move.

Sandwiches and tea we ate in the caves in bad weather, and in good weather out on the rocks. It was surprising how the little robins got to know us, each day they would come right up to us for breadcrumbs, chirping away merrily. Working hard had taken some of the troubles off my mind, I suppose it was not having much time to think. At the end of the day there was the long journey back, preparing the fish, feeding the seals, and then finally a dinner for myself, and afterwards a chat to Mary about how we were progressing. She used to come over at weekends bringing a picnic meal with her, also bringing our dogs Sue and Sweep, two poodles, who enjoyed the run across the fields. While we were building the pools, a contractor was working at full speed on our bungalow, and the Seal Hospital.

Each weekend, although we worked on the pools, we also made a check on how work was progressing on the other buildings. Each day brought us nearer, things seemed to move slowly at first, then suddenly we had all the rock formation in place. Our next job was to build the walls and thirty pillars; another local man helped with these. He was a true Cornishman named Reg; every Friday he had his Cornish pasty, and the only time he stopped working was to light his stinking old pipe. As he lived only one mile away he used to go home to his dinner, making us envious on his return by telling us he had steak and chips, or pasty, or bacon, eggs and beans, and so on. He was a tradesman and did everything right and proper. In the same way as Jack, my mate, he liked everything perfect, even if it took much longer. I was very lucky to have these two good helpers. We worked in all sorts of weather as time was short.

Things were now taking shape, we measured up for the fencing and the manufacturer promised to deliver in two weeks' time. There would be wide gates into each pool so that we could get specially made cages through them, as this would be the only way to move the seals.

Now that the walling and granite pillars were partly completed, the rest couldn't be done until the fencing came, as this had to be tied into each pillar and fitted perfectly. We carried on cementing in the rock formation. This took about two weeks but it was really beginning to look very professional. I was beginning to feel really proud; a few months ago the site was all mud except for the shells of the pools, now something really beautiful was forming, the whole thing blending in nicely with the surroundings and, far from spoiling the area, enhancing it. People walking down to look admitted they had no idea it was going to be so attractive.

The fencing arrived as promised just as we finished the other work, so there was no hold-up at all, we wanted the fencing up so that we could lay the slabs and form the roadways outside the walled area where people would walk.

Our next job was to lay about 500 2ft x 2ft slabs around the pools. This was heavy work as each slab weighed 112 lbs and each slab was set in sand and cement, and many slabs had to be cut as the pools were curved. After laying the slabs, sand and cement had to be placed in between each to seal them, as water would be splashed on them. Also the seals would be moving about on them.

The slabs finished, the firm that constructed the pool shells were notified to come and finish the insides, this being a specialist job. We now had only three weeks to go to the opening; excitement was building up, and the place beginning to look magnificent. I was informed the construction people would be down the following week. So I notified the tarmacadam people to come and finish the road from the Hospital down to and around the pools. About 1000 pieces of turf were laid at the side of the road down by the pools, which previously was all mud. This again was backbreaking work, but everything we did now really helped the finishing touches. The tarmac roads were completed, leaving only the inside of the pools. Only two weeks remained, the Hospital was complete, so we fitted the galvanized pens inside which would accommodate the baby seals. The bungalow was nearly complete and we were promised it would be ready for 27 May – leaving us two days to move our furniture, etc., from St Agnes.

The pool people turned up on Monday. As the large pool was 8 ft deep it took two days to complete the sides, the floor was to be completed last. We were informed that the completed job would be sanded to a polished finish, but this wasn't to be. The firm didn't work weekends so two days were wasted. Only one week to go and two and a half pools to complete!

At this stage my daughter Linda and her husband Paul arrived from the Midlands to help. It took all day Saturday with four of us cleaning the slabs, Paul, Linda, Mary and myself down on our hands and knees scrubbing. Then on Sunday Paul and myself had to get inside the big pool and pumice-stone the sides, because with marcite it is necessary to polish the next day. It seemed we were having to turn our hands to everything, but with only a week to go I couldn't argue too much with the contractors, all I wanted was the job completed. It was like going back to the days of hard labour. The weather was scorching, the site itself was sheltered which made it warm, but inside the pools (which were now pure white) the heat was terrific. We had to work only in shorts. The only pumice stones available were four inches long by two inches wide, and with these we had to rub down all the pools. The object was to remove the top layer of cement, leaving underneath the marcite finish. We sweated profusely, our back and legs were burnt badly by the sun, which made our tasks even harder the following day. Extra men were put on the plastering of the pools, so that one pool was completed in a day, but no extra men were put on to help with the pumice-stoning. People might say I was a fool for doing all this work, but we had to get out of our house at St Agnes by Friday. It was now Wednesday, and one pool to complete and fill with water ready for the seals.

Late Wednesday night we finished rubbing down the last pool and had no skin left on the tips of our fingers.

Thursday we pumped water into the pools, and I must say that although I was tired and sore all over, I was proud of what we had accomplished. After all the problems during the past two years with the opposition, our financial worries, and work worries, here I was standing in the middle of what I think is one of the nicest and most natural layouts for animals of this kind in the country. And I had helped **to** create it myself!

CHAPTER FOURTEEN

The move to Gweek

OUR bungalow was finished, and the removal van had been loaded at St Agnes and brought to Gweek on Friday morning, again with the help of friends and family. All that remained was the really difficult job of bringing the seals over ready to open on Saturday.

I had two special cages made with a sliding door, so that the seals could be transported by the large van (contrary to popular belief seals can be kept dry for quite long periods); as some of the older ones were over seven feet long and weighed between 600/700 lbs, the big problem was going to be tempting them into the cages without damage to themselves or the helpers.

Operation Seal Transportation got under way with Paul and some neighbours ready to carry the cages once they were loaded with seals. Because it would be such a heavy task to lift even one seal in a cage straight out of the pool I decided to try first to get them through the gate that leads to a narrow entrance just outside the pool. Once through that gate the seal would have to enter the cage which would be placed ready in position. It would also have the advantage of our not needing to empty the pool. But everything depended upon tempting each seal into the cage first time for, if we didn't, they would become frightened and we would be in for a very tough time, and nets might be necessary.

First I dealt with the baby seals which, being only a few months old, I was able to carry into the truck. There were six of these, which reduced the number in the pool area, giving more room to manoeuvre the big ones.

I went in the van with the babies, first to reassure them, and also to make sure there was adequate ventilation during the twenty-odd-mile journey. They reacted marvellously, a little scuffling here and there, and an odd cry when one touched the other, but soon they settled down and we were on our way to their new home. Paul drove the van as he had a Heavy Goods Vehicle Licence, taking care to avoid shaking the babies about too much. It took about an hour to get to the pool area, once there we backed the van as near as possible to the gate of the weaning pool, then lifted the shutter at the back of the van and slowly encouraged the seals to that end. I was then able to pick each one up and gently place them at the edge of the pool. One thing I never do with baby seals is to throw them into the pools, I let them get used to the area and atmosphere, then they gently slither into the water without being frightened. The first one in the water sped along the much larger pool, and when the six were in the water, they started jumping in and out, chasing each other around the pool, and leaping about like young dolphins. It gave me a thrill watching them enjoy their new home, but unfortunately I couldn't afford to stay and watch, as the biggest task lay ahead.

We were able to travel much faster on the journey back to St Agnes, taking about forty minutes. A quick drink out of the flask of tea, then the first cage was placed into position by the gate. We checked that the sliding door of the cage worked freely, then two of us went inside; one to operate the gate, the other to tempt the seal into the cage. I had deliberately not fed the seals that morning in the hope that, being hungry, they would go after the fish we were using as bait inside the cage. The first to jump out of the pool was Benny, he weighed over 500 lbs. I threw the fish as far back into the cage as I could, he moved slowly towards it, edged half his body on to the cage floor, then changed his mind. I crawled into the cage and moved the fish a little farther back, he eased himself in again about half way, so we tried sliding the door down gently on his back thinking he might rush forward, but instead he backed himself out and jumped into the pool. He would now be a problem. Flipper and Benny usually jumped out to me when I went into the pool area, but this time only Flipper jumped out; he was older and bigger than Benny, I gave him a pat, threw another fish into the cage and to my delighted amazement he went straight in after it (he always was the one with the biggest appetite), so quickly we pulled the sliding door down – that was the largest seal safely caged up! It took six of us to carry Flipper from the gate to the van about twenty-five yards away. Left in the pool we had Benny, Jenny, Sheba, Kim, Alpha and Sixpence. We knew we would have a problem with Kim, the sea-lion, so we were leaving him until last. Sea-lions move very fast, and can run on their four flippers.

We didn't want to frighten or damage the seals during this operation, however long it took. It was six o'clock in the evening when by trial and error, and crafty movements, we had managed to tempt all the other seals one by one into the cages. Sometimes it took three or four attempts but leaving the morning feed had helped to do the trick. The last one in was Alpha, more cautious than the others, but finally he too was in, leaving just Kim to catch. As he couldn't be tempted in, we decided to empty the pool.

While this was emptying we made another journey to Gweek with the six seals already in the van, two in the cages and four loose in the van. Again I sat in the back to reassure them that all was well, and they were on their way to their new home.

These seals were to go into the residents' pool, so we backed the van to the gate, opened the door and as the drop of the van was low, put a big board for them to flap down on, and straight into the pool. We opened the cage doors to let out the other two, and soon they were all happily swimming in the pools, their heads popping up and down taking in all their new sur-roundings. It was now nine p.m., so we had a quick sandwich and cup of tea, before the last journey; reaching St Agnes at 10.30 p.m. We were desperately tired, but we had to make the final effort. We didn't know how we would tackle Kim; he was big and fast, he could climb in and out of the pool making things very difficult.

We placed the cage at the edge of the pool; Kim was lying down on the bottom looking really miserable without his mates. I spoke to him and his eyes kept blinking as if he understood. I went towards him, he got up and climbed out of the pool towards an opening of one of the pens we kept the babies in. We quickly moved the cage across the opening blocking him in. At least we had him trapped, but not yet in the cage. He went into the babies' pen, so I climbed the wire and went towards him and tapped his bottom gently. To my amazement he turned and went straight into the cage. Paul quickly slid the door down and that was it. It was midnight when we unloaded Kim at Gweek, and watched his excitement when he saw all the other seals playing in the pool. Now for a good night's sleep before the unofficial opening in the morning to the local residents.

CHAPTER FIFTEEN

The grand opening

THE great day had come; years of planning and hard work had brought us to Whitsun, 1975. A new and large Sanctuary was ready to be opened to the public. Over £40,000 had been borrowed to complete the Sanctuary, running costs were going to be high, so an entrance charge was necessary if the Sanctuary was to survive. We had plenty of publicity due to our opposers, so we hoped that people would now come to see for themselves what we had actually done, and make their own decision as to whether or not we had spoiled the area.

We didn't have much sleep Friday night, or rather Saturday morning. We were up at seven o'clock and down at the Sanctuary at eight checking that all was well, and putting in any finishing touches. Linda took over the pay desk, Paul saw to the filters, Mary went into the Hospital, and that left me to meet people, help Paul with the filters, prepare the fish and feed the seals at 11.00 a.m. and 4.30 p.m. Saturday was an open day for the residents of Gweek and surrounding areas, admission was free, so no money was taken at the pay desk, just information given to those entering. The idea was for all the local people, those for and against the Sanctuary, to come and see the completed work, and give any criticism they might have.

All who came were overwhelmed by what they saw, and I had nothing but congratulations. I felt happy and proud. Everybody watched the seals swim and play in their new environment, really enjoying themselves in the large pools.

Even feeding time was different for me, as I had a larger area of pool to throw the fish and could see the seals really move for the first time, chasing the fish. Little talks were given to people who wanted to know more about the seals, in all the day went off very well. I kept walking up and down the pools talking, and although I was really tired after the hectic weeks, somehow I found the energy to make this opening day one to be really proud of.

I don't exactly know how many people came but I do know all enjoyed the visit. Now we were to see how many would come on the official opening the next day, Whit Sunday. Although we enjoyed the Saturday and talked ourselves hoarse, we were glad to get home and have a good night's sleep. The house wasn't straight but at least the beds were made. We discussed the day, had a bit of supper and went to bed, praying the Sanctuary would be a success. During all this time of anxiety, the planning, the opposition, the construction, Mary and myself had never talked a lot. I had tried not to worry her too much, but naturally she was worried over what had gone on; also I think for my health, as a person can endure only so much. However, with the worst past we now had to look forward to the future and meet any troubles as they came.

It didn't take long getting off to sleep that night, and when the alarm went in the morning I felt I had only just put my head down. Anyhow up I got, made the tea and called Mary. We got ourselves ready, loaded what we wanted into the Land-Rover including sandwiches and flask of tea, then were off around the road through Gweek village and around to the car park. We had a small wooden kiosk for giving tickets to the visitors; Linda, my daughter, was taking charge of that and Paul would be helping with the filters. We opened the gates leading up through the woods to the Hospital, and drove down to the pools. It was a beautiful morning, the sun was shining, the birds were chirping, and an odd rabbit ran across the road, as if everything was greeting us on our big day. As we approached the pools, all the seals jumped out as if full of excitement themselves; as we drove the Land-Rover past each pool, the seals flapped along the side following us up. I shouted to each of them as I passed, I felt good that morning and was looking forward to welcoming visitors. We did the filters, cleaned round the sides of the pools, then we were ready. Ten o'clock was the opening time; at ten past ten no one had come, at ten twenty-five, two people came down the hill. These were our first paying customers after eighteen years. I greeted the young couple, explained each pool to them, what was wrong with the seals, then let them walk along the pools. The seals put a show on for them by chasing each other in and out of the water, and where the slabs had got wet they slithered along like children going down a slide. A few more people followed, so I went up to the Hospital and prepared the fish for feeding at 11.00 a.m. When I got back down to the pools with the fish there were about 20 people.

I wasn't too disappointed, as Sunday is a funny day: people prepare dinners, go to church, so one couldn't expect too much for the first day of the holiday. I fed the seals, creating as much splashing as possible, getting the seals to speed up the pool after the fish. Between each pool I chatted to the crowd explaining the difficulty of feeding some of the seals. It was the younger ones that had this difficulty. There was Silky with one eye, Nelson also had one eye (he had been stoned on the beach by teenagers, losing his eye and sustaining concussion). Another seal had smashed its jaw, lost part of its flipper, and had ulcers all over its body. These took

longer to feed, but the feeding was entertaining in its
own way. All the visitors enjoyed themselves and took
some good photographs. Throughout the country there
must be hundreds of photographs of our seals over the
years, now with layout of the rock formation and
caves, the beautiful surrounding countryside, people
would be able to get really magnificent pictures. Up
until 2 p.m. about 100 people had been through the
gates, then suddenly they came in masses. We were a
success. I noticed that the majority of people were
Cornish, and this made me happy, for to get a lot of
local people interested spelt good publicity; if they enjoyed it then they would recommend others to come.
Also the seals were part of their heritage, and it was
good to know they were taking such an interest in our
work. I circulated amongst the visitors as much as I
could, being congratulated on beating the opposition,
and also how well the work had been done.

Monday was different altogether. From 9.30 a.m. onwards cars poured into the car park, and by 11.30 a.m.
the car park was full, cars were double-parked along
the creek, right out through Gweek, up the hill by the
chapel, in fact the police had a job getting through.
The main reason for this was that I had been limited
to 30 cars in my car park, and therefore they were
parking wherever they could. At least it proved the
public were interested. It was a mixed crowd, some
locals, some visitors. I had written notices by each
pool, also general information as to what our work
was about, and how we did it. The seals seemed to
enjoy the crowds, leaping about, and at feed time
splashed us all by jumping in and out. I was asked
many questions about our work, and also about the
opposition we had endured. As crowds went out, so
more crowds came in. Whit Monday and Tuesday
were frantically busy, but we enjoyed every minute

though it was exhausting. At least the Sanctuary was launched, and people were coming to support it. The main complaint was that we didn't supply drinks or ices, and it was difficult to explain that the planning authorities wouldn't allow us to. It is quite a walk from the car park, a beautiful walk but on a hot day it makes one thirsty. Maybe one day the planners will give in to the request of the majority rather than the select and powerful few. The main thing was that we had plenty of room for the seals, and also room for people to view them properly and also to walk round and really enjoy themselves.

We didn't have a completely trouble-free opening, since a few of the new people that had just moved in to Gweek did try on a few nasty ploys hoping to turn away visitors. We also had to run a mini-bus for the very old, crippled and people with heart or lung diseases, otherwise these people would never see the seals, and a small charge had to be made just to cover the petrol expenses.

What did surprise me was that although we had the mass crowds, the place was very quiet, and other than cigarette ends, there was no litter at all. Everybody respected what we were doing, children were prevented from climbing the fencing and were made to pick up any toffee papers they threw down. Unfortunately by the end of March all sick seals finish their Hospital treatment and go into the pool, so there was nothing to see at the Hospital except the layout.

The seals were spread out in the pools, with the old ones in the residents' pool, Silky with one eye, Blackie with damaged jaw and bad lungs in the Isolation Pool, plus Nelson who had lost an eye, one with part of its flipper lost also, with a damaged jaw and some nasty ulcers under its body, and two females again with damaged jaws.

A friend of mine needed some grazing for his mare, Toh-Toh, who was in foal. She stood 14.2 hands, was quiet and broken, and loved a bit of fuss. She was due to foal in a few days, and had been grazing with me most of the winter. Some people I knew were getting rid of their horses, and I took a Shetland pony to save it being put down, as company for Toh-Toh. It was called Little Blue, as it had blue eyes. It was very nervous, as if it had been treated roughly. Once you had caught it, you could do anything with it, but catching it was the problem. It was a stocky little pony, too fat if anything after grazing over twenty acres of grassland. Every time the ponies saw me they would call; naturally they wanted their lumps of sugar, and this became a daily habit.

It was the end of June when Toh-Toh gave birth to a bay-mare foal. It was a scorching day, the sun over-head, so immediately I called her Sunshine. She soon got to know me, and soon started getting fuss from the visitors. It was surprising how quickly from being a weak-legged foal she was tearing around the field teas-ing her mother. Little Blue took it all with good grace, and the three followed each other from field to field.

In August the RSPCA brought me a grey Dartmoor pony called Pippa. He was a two-year-old stallion, halter-broken but that was all. I had now acquired quite a family to care for with seals and horses. Pippa was a little terror. He used to run up behind you and bite your bottom, and Mary would walk half a mile more around the field to get out of his way. It was only devilment, but he knew he had her scared and played on it; I used to tap his nose, then he'd stop.

Each morning as we prepared the Sanctuary the ponies would all come trampling down to the fence waiting patiently for their sugars. Toh-Toh used to hang her tongue out after having her sugars and suck just like a baby.

The summer was going well, we were getting the crowds we expected, the days were long, 8 o'clock in the morning to ten o'clock at night. Linda, my daughter, used to go at six because of her children, she had two daughters aged one and two years, Lizzy and Teresa. During the day the eldest, Lizzy, used to walk around the Sanctuary chatting to people, telling them about her granddad's seals – she captivated as big an audience and had as many photographs taken as the seals.

Paul, my son-in-law, worked most of the time with me. People at the Sanctuary used to envy us our jobs, because of the work we did and the beautiful surroundings we worked in, but it is a job of very long hours, seven days a week, every week of the year, which would not suit everybody.

During the season we met people from all walks of life, and from various parts of the world. Nearly all the people were interested in the work we did and fascinated by the seals. By meeting new people and talking personally to them, the days didn't seem quite so long, also it was interesting, as we met doctors, vets and oceanologists, as well as the general public.

September approached, people were still coming to the Sanctuary, but now we had to prepare the Hospital ready for the arrival of the first baby. People who had visited us earlier in the year were ringing up to find out if the first baby had arrived. On 21 September we had a call about a baby seal, from Portreath. It was only a few hours old, its creamy white coat was like silk, the eyes were large and shiny, there was no damage, so we put it gently into the Land-Rover and drove back to the Hospital. On arrival people were waiting for us to return, as most had never seen a baby seal. When I opened the flaps at the back of the Land-Rover, their faces lit up, and there were the little smiles and cries of

'Oh isn't it beautiful'. I carried it up into the Hospital, settled it down into its pen, and then set about making its first feed with the bottle. I skinned three mackerel, boned them, then emulsified the fish to a liquid, adding warm water to bring to blood temperature. Vitamins and oils were added. Its bottle now ready, I tried to get it to suck. I placed the teat into its mouth but it pulled its head away. I squeezed a little milk into the side of its mouth, to give it the taste, then tried the teat in again. It clenched its teeth on the teat, and pulled hard, but did not suck. I kept repeating this, small quantities went down its throat, but it took about twenty minutes before I got it sucking, then the bottle didn't last long. When a baby seal suckles its tongue comes out and makes a semi-circle, and the suction is very powerful, it will take four bottles of fluid quicker than a human baby will take one bottle. The mixture in the bottle is thick, just fluid enough to pass through the large hole in the teat.

CHAPTER SIXTEEN

The winter of 1975-6

TWENTY baby seals were washed in during our first winter at Gweek, their ages ranging from one day to six weeks old. Some of the younger ones were in good condition, and after six weeks' attention were able to return to sea. A few of the older ones were starved and had minor injuries, and these took a little longer to get fit so that they could cope with the rough seas they would be returning to. Two seals had major problems, one that was named Lucky, and one we called Spitfire.

Lucky's name was given to him by the Benson School at Oxford. The children there adopted him, and chose his name. By adoption I mean the seal stays at the Sanctuary, but the children collect money to help to feed him. Then they visit him occasionally, or the teachers come down and take back photographs, and we write regularly about his progress. We have two adoptions at the moment, the other one by some office girls who work in a Midlands factory.

Lucky was washed up at Godrevy, near Hoyle and when I went to pick him up I thought he was a walrus type of seal. His face looked as if it had been bashed in at the front, he had no nostrils, no front teeth, his breathing was queer, his ears were badly infected and he was very thin. After a close look we saw that he had a cleft palate, and a hare lip. Feeding was to prove very difficult, as he made no attempt to take anything himself, in fact, he showed no interest at all in fish; from our experience it seems that seals after being left by their mothers very rarely learn to catch fish until they reach starvation point. Unfortunately very often at this time, gales blow up, seas roughen and as they are very weak through the starvation, they get smashed against the rocks. Lucky had no serious injuries, and he must have been suckled by his mother in some fashion to have survived. He was about three weeks old, so if his mother hadn't fed him he would have died in the first week. We had six other baby seals at the time, so I put Lucky in with another seal, a female about the same age who was taking fish quite well which I thought might help to encourage Lucky. I sorted out some very small mackerel, firm ones that we could push to the back of his throat to make him swallow. It takes two to force feed a seal, one holding the seal (which is always me) forcing open the jaws so that the other could push the fish down. Trying to open his mouth was a funny sensation. He had no full lip at the top like other seals, the mouth was very wide and didn't join up to form proper nostrils, the breathing holes seemed to be in the top palate, my fingers kept slipping, and I seemed to lose them in the soft tissues of the lip. Finally I was able to get a grip and opened the mouth enough for Mary to slip the fish in. His great big eyes looked right into my face as the fish slipped down into his stomach, I suppose it was a funny sensation. We never give too many fish at first as this is their first solid food, and we need to get their digestive systems used to it gradually. Very often when babies are weaned on to fish, they suffer badly with constipation, so a little liquid paraffin comes in handy. Lucky's ears were weeping badly, he was given injections, also a special cream squeezed into the ear, which is only a tiny hole just behind the eyes. He was given four feeds a day for the first week, then cut down to three, and finally two feeds a day. His ears were quite a problem, a specimen was sent away, to find the antibiotic suited

to that particular virus. Children as well as animals born
with a cleft palate suffer the same thing with ear infections. My other worry was whether he would be able
to swim in the pool, not being able to close his nostrils
when diving like other seals. Also whether he would be
able to swim, and sleep, under water as other seals do.
After feeding Lucky in the pen he used to nuzzle up
to his female partner, being a gentle soul, and if the
female happened to growl at him he'd look up at me
sadly as if to say, 'I'm a nice-looking guy, why is she
cross with me?' for as far as he was concerned he was as
handsome as his mates in the next pen. One couldn't
help but keep looking at him, for he was such a strange
sight. I would like to mention here that any jokes I
might make about Lucky are not meant to be taken
seriously, as I know many human beings too with
cleft palates. I think the world of Lucky, and it is not my
nature to laugh at any deformity. Hundreds of spastic
children visit my Sanctuary and I try to give extra time
to them. But I feel a sense of humour always helps
as long as it is not offensive.

Lucky was in hospital for six weeks, his one ear had
dried up nicely, but the other still wept a little, however. I thought I had better try him in the water and see
what his reaction would be. He had put on quite a bit
of weight since he first came in, so rather than carry
him to the Land-Rover, we used a small cage built
specially for the purpose, and as he was used to his
female companion, we took both down to the small
pool, ideal for weaning, i.e. learning to pick up the
fish for themselves. The female had no hesitation in
going into the water, but Lucky was reluctant. He pushed
his nose over the side of the pool, as if he was testing
the temperature, then pulled it out and started flipping
around the edge of the pool. I walked over to him and
stroked him, then with a gentle push I heaved him

into the water. At first he did a sort of dog's paddle, keeping his head above water. The female swam circles around him, as if to say, 'you can't do what I can do'. As if he understood, he dived sharply after her chasing her round the pool. After short intervals he came up breathing heavily and even when under water bubbles came up showing something was slightly wrong. With his disability it was obvious that he would have to adapt himself to the water differently from the normal seal. At feeding time I tried throwing a few fish into the water, but neither Lucky nor the female took any notice. I enticed them out on to the side of the pool and the female took the fish well from my hand and soon had her tummy full, Lucky waiting patiently on my other side. I held the fish just above his head; he opened his mouth which looked twice the size of the female's. Looking inside I could see the two holes that should have been part of his nostrils, I wondered how he would keep the water out of his nostrils when he was under water. I lowered the fish gently into his mouth holding the tail of the fish, by the time it was going down his throat, my fingers seemed to be up his flabby nose. He soon took fourteen fish, then slithered back into the pool. At least he had settled down to being in the pool, and I knew that often seals didn't eat the first day we put them in the water. Gradually he seemed to stay under water a little longer each time but did a lot of spluttering when he came up. The female used to tease him, so he chased her round the pool, when he caught up with her and tried to bite her round the neck as seals do, he found he couldn't because he had no front teeth. Occasionally we had to catch him to put cream into the infected ear, he didn't like this one bit, and he was now getting very heavy to hold. We knew Lucky would never go back to sea, so we could study his behaviour, in and out of the water.

Each week that went by he seemed to stay longer under water. For six months I fed him by hand, then suddenly one day he picked a fish from the water, and after that he would only take them in the water. Now he was able to take fish properly in the water, we decided to put both Lucky and the female into the larger weaning pool – over sixty feet long – with four other seals including Nelson, the one that had been stoned by teenagers on a beach at Plymouth.

Lucky got on quite well with his new friends, the only trouble was that when I threw him a fish the others could take it out of his mouth, as he had no teeth to grip it with. It was from this time that I brought in the little humour in my talks to the visitors telling them that Lucky had a cleft palate and hare lip, but not to laugh at him, as he didn't know, the other seals wouldn't tell him, I wouldn't tell him, so they mustn't.

Spitfire, our other invalid, was washed up on a slipway at Portreath. It was nearly dark when Paul and I got him into the Land-Rover. We could see he was suffering from malnutrition, his eyes were badly infected, and his breathing was bad. He had an injection on arrival at the Hospital, his eyes were bathed and he was left to rest for the night. In the morning he would be force fed with small fish.

When I opened the pen door, he spat furiously at me, but I had had this many times over the years, and thought I knew how to cope with him. I managed to grab him and get him held tight between my knees; I placed my fingers each side of his jaw to open his mouth, but the harder I squeezed the tighter he held his jaws closed. Suddenly he started writhing and bringing up his back flippers, until finally he twisted himself from my grip towards my back, and as I went to move he bit my bottom. Over a period of a week he bit my

Simon II asks Sally II to come into the pool

The sea-lions Kim and Judy

Oiled guillemots

Blacky's eyes and body showing burn marks

Baby seal just put into the weaning pool
Benny and Jenny

Playtime never ends for seals

Jenny with her week-old pup suckling
Spitfire

THE WINTER OF 1975-6

bottom six times. He was a very difficult seal to feed, and although we managed to get some fish down him it was not the amount of fish I would have wished him to have. He was very thin and needed a lot of fattening up. He exhausted me every feed time, until finally I decided to take him down to the pool and see if he might take the fish better in the water. We used the small pool; the first day he wasn't interested, but the second day he started playing with the fish in the water. Very often I cut a fish in half as some seals eat it better this way. When he jumped out of the pool, I held out half a fish, the tail end, and as he opened his mouth, I tried to push it in, but he rejected it. I tried again, but this time I just held it to his mouth, he pushed his head towards the fish and took it. I repeated this a few times and he took it each time. The tail end of the fish is soft to the palate, he seemed to prefer this, and ever since he eats only the tail ends. Even now if I throw him a whole fish, he catches it crossways in his mouth, and with his flipper breaks off the head. At feed times now when he jumps out at my feet I tell the people that this is Spitfire, and that every time I tried to feed him at the Hospital he bit my bottom so that I'm the only white man with a black bottom. Then I ignore him, throwing the fish to the other seals as I walk up and down the pool. He follows on my heels, but I pretend he's not there. I can hear the people saying, 'Oh, give him some'. I keep going up and down; he looks pitifully at me, so then I ask the public if I should give him one, and naturally they all chorus 'Yes'.

So I throw him about ten half-fish one after the other, then after feeding the others again I give Spitfire some whole fish which he breaks up, in doing this he rolls over and splashes into the water. He seems to do this regularly, as if playing to the crowd.

Every seal has a character of its own, not one is the same as the other. One of the pups that was born to us that winter we called Terry. He has a beautiful nature, and always comes to me when I enter the pool area just as Simon did. I kept Terry in the small pool for a while, and when people ask me if I breed with the seals, I turn to Terry and then to the public and say, 'Yes, don't you think he looks like me?'

I had to give time too to some of the other seals, especially Silky (she only had one eye), who looked forward to her ten minutes' playtime. When I first go to see her (she is in the isolation pool) she comes near to the side of the pool, and I have to hold her flipper, and tickle her tummy. Then I pretend to go, and she darts out slithering along the wet slabs, and grabs my trousers. When she releases them (I have to wait for that, or she'd pull my trousers off), I run around the pool; first she chases me, then seeing she cannot catch me, she dives into the water, taking the short cut, and ends up in front of me. Seals can swim very fast so it's no effort to beat me.

Money was very tight throughout that first winter at Gweek. Hanging over my head was the repayment of the £43,000 bank loan at an interest rate which was increasing all the time. I knew that I was in for many tough years before this was paid off. More immediate was the struggle to keep abreast with the running costs of the Sanctuary – the bills for fish and electricity were particularly heavy and some of my cheques were stopped by the bank because I had reached my maximum overdraft. As usual I found relief from my worries in hard physical work. There was much to be done preparing the Sanctuary for the start of the next visiting season in the spring. First the Hospital needed a really good cleaning and disinfecting for, although it was washed down regularly, the natural odour from the seals hung around. But some vigorous scrubbing with sweet-smelling disinfectant took care of most of that. All the boards were taken out and scrubbed, the walls washed down and painted and the Hospital looked like new again. Rain had washed soil on to the long drive-way and piles of leaves had built up all along the road leading to the Sanctuary. We cleared this all away and lopped off overgrown branches from the trees so that vehicles would be able to have a free passage.

All this work had to be done in intervals between feeding and treating the sick and injured seals. But thankfully it had been a relatively quiet winter and seventeen of the babies rescued were eventually fit enough to be returned to the sea, leaving Lucky, Spitfire and Terry to join our long-term residents. A few injured birds had been brought in but these soon recovered and were able to fly off again. Everything was in order for the new season and we looked forward to the time when the visitors would start coming again and, please God, a little money would start trickling through to the bank.

CHAPTER SEVENTEEN

Feeding time

ON a beautiful summer's day when the surrounding countryside is at its best and the seals are disporting themselves happily in the pools, visitors often remark on how much they would like to exchange their lives in factories and offices for mine. I doubt if many of them really would when they came to realize that it is a seven-days-a-week job with no holidays. There is nothing that I would rather be doing in the world than looking after my seals but it is hard physical work which doesn't leave much time for relaxation. A typical day starts at 7.00 a.m. Sleepily I switch off the alarm and pop out of bed before I can succumb to the temptation of having another five minutes. After washing and dressing I fill the kettle and while it is boiling I put milk and sugar in the cup and in the flask I will be taking to work. My breakfast is a Shredded Wheat with cold milk, a cup of tea and a cigarette.

At 8 o'clock I take my basket with my sandwiches, put the flask in, pick up from the kitchen any apple and potato peelings to take down for the ponies and donkeys, and check if I have any sugar cubes for them. Then on with my wellington boots and jacket, a quick check outside at the fishpond, feed the fish, then over to the Land-Rover parked by the shed outside the house.

In the shed I store a special powder, J100, which I use in my filtration plant. At least one bag a day is used, sometimes two, so the powder is purchased in bulk about 500 bags at a time. Each bag weighs only 28 lbs which is light considering the size of the bags, about 1½ times the size of a bag of coal. I put at least one bag in the back of the Land-Rover, and then commence the drive to the Sanctuary.

My house is built in the grounds of the Sanctuary, and although it is only a few minutes' walk across the 12-acre field to the Hospital, I have to drive down to Gweek village, along the creek to the car park, in order to open the gates for my staff or visitors.

The road along by the creek is full of pot-holes so I drive slowly, and glance at the swans and sea birds on the water. The tide is in, which alters the look of the creek, making it one large lake. When the tide goes out, there's nothing but mud and old trees that have fallen from the banks of the creek, rotting away over the years. In the winter there are 70 – 80 boats stored on the boatyard, resting after their busy summer taking their owners fishing or pleasure cruising, all sizes from rowing boats to large cabin cruisers. About 200 yards along the creek I turn into the car park, then on to a 10 ft tarmac drive about 400 yards long cut through the woods, which makes a beautiful walk for the visitors. About 100 yards up the drive is a gate which is locked every night, I open the gate and drive to the end of the road through the woods at the end of which is the Hospital. I take my basket with my sandwiches into the office adjoining the Hospital, have a quick check on the baby seals in the pens, wish a good morning to each one, then back to the Land-Rover for a short drive of 200 yards down a steep slope to the pools. This short drive takes in one of the loveliest views anyone would wish to see – the creek winding its way down towards the sea surrounded by woodlands and sloping green fields, with cattle grazing happily on the slightly frosted grass. I think of all the people who live in large cities and work indoors all day, also of the time when I worked in the dark dusty mines of

Wales and the Midlands. It's always a thrill to start the day, with the fresh, unpolluted air filling my lungs, and the gorgeous views which vary according to the seasons. When the sun shines there isn't a prettier part of Cornwall. Having lived for eighteen years at St Agnes overlooking the sea, watching the massive waves pound the cliffs, Gweek is so restful, with only the birds chirping away, or the sound of a boat chugging its way up the creek, to break the calm stillness.

The road leading to the five pools splits like a letter Y, one arm leads through the pools splitting the isolation pool and the residents' pool from the weaning pool and the two small breeding pools. The other splits to the end of the breeding pools, both roads joining up with one long road fronting the five pools; this is the viewing area for the public.

I stop the Land-Rover by the breeding pools and then check each pool in turn, seeing first that all the seals are OK, and then examine the flow of water into the pools. This tells me the condition of the filters after circulating all night.

I talk to the seals, wishing good morning to each one. Most of them jump out of the water to give their greetings, sometimes giving me a good spraying at the same time. Those who do not jump out sit up in the water stretching their necks as I move on to the next pool and following me until I'm out of sight.

Once my check is complete and all is well, I then make for the filter house built below ground near the pools. There are four pumps serving five filters. Each filter has a pressure gauge, which tells the state of the filters. When clean the pressure reads 15 lbs per square inch, anything over 22 lbs shows that the filters need cleaning. and naturally after circulating water from the pools all night, they need doing first thing each morning and again in the evening.

The filters are called ditremetus earth filters. These are stainless steel containers, about 3 feet 6 inches high, and 2 feet in diameter, having a dome-shaped lid held on by a stainless steel ring pulled tight by a clip, as the pressure of water inside is terrific. Inside each container are eight half-moon plastic ribs 2 feet 6 inches high and about 14 inches wide, covered by a special cloth which allows water through under pressure. The water goes in one way, that is through the outside of the pads, and flows into pipes back to the pools from the inside of the pads. A special powder has to be used to coat the pads acting as a filter, all the dirt and grease sticks to the powder allowing only clean water through the pads and back to the pool. It is all this dirt collecting on the powder that gradually sends the pressure up on the gauge, so twice daily this powder has to be changed to obtain clean pools and keep the filters in good order. Sometimes the cloth pads tear and have to be sewn up, or one of the plastic ribs breaks and has to be repaired.

The powder is coated on to the pads by first dissolving it in water, and then it is sucked in through a pipe which evenly coats each pad inside the filter. Before the new powder is put in certain valves have to be closed or opened to backwash the filters. This is reversing the flow of water through the filter so that it cleans off all the dirty powder and washes it through a waste pipe to the drains. This means that each day water has to be wasted to clean the filters, so each pool has to be topped up occasionally to fill the pool to its original level. The water is sucked in from the creek, and is sea water.

Once each filter has been cleaned, the pools are swept, all the sides washed down, and a little chlorine added to keep the water free from bacteria. It means that each pool is kept like a swimming pool, except that with seals there is a lot more waste material deposited in the water, so there is more work involved. Special equipment with the use of tablets tells us the state of the water, but luckily we are able to change the water completely every two or three weeks.

When all the pools are clean I then go and have a quick chat with the donkeys and ponies; they look forward to their sugar lumps and titbits. I see that they have sufficient water, then back up to the Hospital. On arrival it's like entering a maternity ward, with the babies crying away for their breakfast. A few comforting words help to reassure them it won't be long. The cry of a baby seal sounds very sad, it's also quite deafening when there's a lot of them crying away together. Some of the babies are on fish, but have to be force fed, this takes two of us, one holding the seal, the other pushing the fish down. The others are still suckling and luckily they take it from the bottle.

Mary now joins me for the feeding, so we first feed the ones that take fish as this is the hardest part. No one likes to be force fed, not even a seal. The average feed is about eight mackerel, except when there is damage to the head, or the mouth is swollen badly, then we give about four mackerel five times a day. The youngest resist the feeding but gradually, with a bit of patience, the mackerel given whole slither down out of sight. Sometimes the excitement of feeding causes the seal to wet itself, and naturally the one who is holding gets the brunt of it. As each baby has its fill, it rolls over, contented to feel its stomach full, and goes to sleep.

For the bottle feeding sixteen mackerel have to be skinned, boned, and placed into an emulsifier, with distilled water to blend it into a semi-liquid. Vitamins and antibiotics are added, and well mixed up. The mixture, thin enough to pass through a large hole in a babies' teat, is brought to blood temperature. The mixture is poured into two-pint jugs, sixteen mackerel yielding about four pints. If the mixture is too thick, then a little more warm water is added to give it a nice consistency, but not watery.

Once the baby starts sucking the two pints are soon gone. Babies from ten days old have four feeds a day. After feeding all the utensils have to be thoroughly cleaned, the skin and bones of the fish disposed of, but all this is done in peace and quiet. All crying from the babies has stopped, but there is still much to be done; twenty seals down at the pools are waiting anxiously for their food. All fish is defrosted overnight, unless fresh mackerel has come. This has to be sorted into five buckets, graded according to size, as some of the fully matured seals take the large fish, some of the ones with damaged jaws eat half-fish, younger ones eat small fish. Each pool has its own bucket, so work begins in sorting it out. When completed, frozen fish is placed in a large bin to defrost for the evening meal. Small fish for the babies is done separately as they need more feeds. About fourteen stone of fish is used daily.

The time is now 10.45 a.m., so a cup of tea is welcome before feeding the seals down at the pools at 11 a.m.

Sometimes the routine is altered, as we may get a telephone call about a baby seal washed up on the beach. After the morning cuppa and a quick look at the daily paper, the buckets of fish are carried to the Land-Rover just outside the Hospital. A 200-yard drive down the hill brings me to the pools, and as I drive past each pool in turn, out jump the seals. I usually drive to the large pool first and make my way back along each pool.

At the time of writing the residents of the large

pool are Kim, Benny, Flipper, Sheba and Jenny. Nearly all weigh between 600 and 700 lbs, their fur coats are shabby, as this is the period of the year they moult, and a new fur coat is grown by March. While moulting they spend a little more time on land, and sometimes go off their food. It's not the case at the moment, however, they are all waiting anxiously for their fish. Jenny and Benny jump out of the pool, the bulk of their bodies creating a huge splash spraying me even outside the pool. As I walk the twenty yards to the gate, the other seals in the water create massive waves in the pool by rushing from one end to the place where I enter the gate. As I get to the gate Jenny and Benny are flapping along the side of the pool, and the other three fly out of the pool splashing water everywhere. I usually carry two buckets of fish, one and a half buckets for the large pool, and half a bucket for the isolation pool next to it. I am only half way through the gate when Kim the sea-lion, who is very fast, has his head in the bucket, whips two fish and dives into the pool. In the meantime Benny has reached the gate; I have to place the two buckets down on the floor to shut the gate, and no sooner have I done this than Benny is helping himself. Once the gate is shut the fun and games start, fish and seals flying everywhere. I try to see that each one has its share, but occasionally Kim, who is twice as fast as the others, and can leap into the air like a dolphin, gets a few more than he should.

If I miss giving Flipper the right number of fish, I am soon told off. He is the only seal who talks, in his own language of course, but it's great fun listening to him. Towards the end of the feed he continually asks for more fish, and as I open the gate into the next pool, he jumps out, so I have to close it quickly after me, but his talking continues. I throw him a fish to shut

him up, and start feeding the three in the isolation pool.
These are Silky, with one eye, Blacky with a broken
jaw and lung congestion and Susan with a damaged
jaw.

After the speedy feeding of the large seals, the three
in the isolation pool eat very slowly. Their fish have to
be cut in half, because of their damaged jaws, and even
then one will hold the fish in its mouth and let one of
the others bite half of it off. A few more fish are thrown
to Flipper, still talking away merrily. Visitors love to
hear him.

The buckets now empty, I place these in the back of
the Land-Rover, and take out two more for the weaning
pool. Again as I walk to the gate eight seals come tearing down the pool. These are much younger and smaller
than the large ones, but they have the same appetites.
Their ages vary from one year old to three years old,
and all have various injuries. The weaning pool, as we
call it, was first used solely for that purpose, in teaching
the baby seals to catch their own fish. But now we
have built two small pools specially for the babies.

In the weaning pool we have Nelson (one eye), Spitfire (eyes and jaw damaged), Terry, Lucky (cleft palate),
Tubby, Cheeky, Footy and Streaky. Terry is the most
awkward to feed, I still have to push the fish right
down his throat, and if the head of the fish is in any
way torn he will reject it. Every time I start to feed
Terry, Cheeky jumps out of the water and keeps nudg-
ing the bucket. Lucky follows quivering his top lip,
which just sags all in one, while the normal seal has
nostrils. Sometimes I hand Lucky a fish, but Cheeky
steals it from his mouth. The deformity of his mouth
makes it impossible for him to hold or tear a fish, he
just has to swallow it quickly.

To get rid of Cheeky while I concentrate on Terry, I

throw a fish along the side of the pool, and she scampers off, very often losing it, as the other seals in the pool beat her to it. I never rush Terry with his food, he takes it in his own time, and I just have to wait. Once Terry is satisfied, I take the last bucket to the new babies in the two small pools. All are between 12 – 14 weeks old, most having spent 8 – 10 weeks in the Hospital, now they are learning to catch their own fish.

All the babies meet me at the gate, and as the gate opens inwards, it takes some time to push them out of the way. Once in you need eyes at the back of your head, all their mouths open as each fish is taken from the bucket, so I have to be quick and careful that in feeding one, the others don't snap at the fish and bite my hand instead.

An odd fish is thrown into the water for them to try and pick up. Some learn quickly, some are slow catching on. The same applies to feeding by hand; some have terrific suction, with others you have to push the fish all the way down, my fingers practically touching their teeth on the last bit. The reason for this is that they haven't learned to use their tongues properly, but in time they all get the idea.

The last three babies are a little more awkward to feed; two of these were washed in on the same day only a few hours old, and were very difficult to wean on to fish. The only way they will take it, is by lying on their backs with me pushing the fish right down. The other seal had a very badly damaged jaw, so I feed him a little slower, and away from the other two, to avoid disturbance and help concentration. I talk to all the babies fed by hand, as this seems to take away their fear and usually gets them eating.

The feeding now completed, back to the Land-Rover

for the return to the Hospital, where I wash the buckets. As most of the fish I purchase are frozen, I get out enough fish for the afternoon feed, filling a large plastic bin, which holds about seven stone of fish, then warm water is poured in so that the fish defrosts by about 3.30 p.m. I also get out other fish and defrost this in a bucket, as this will be needed for the babies' feed about 2.00 p.m.

By 11.45 a.m. Mary has cleaned the Hospital and office, so I now give her a hand to clean out the seal pens. Like all animals, and children, seals pass waste products from the body, so cleaning daily is necessary. There is a very strong ammonia smell from the waste matter of a seal enclosed in a building, so a good scrub down with disinfectant is needed, firstly the slatted boards that they lie on, then by lifting the boards the floor beneath.

By 12.30 this is done so we decide to have a cup of tea and our sandwiches. If there are no visitors (we get very few in the winter) Mary goes home in between the feeding of the baby seals, as she has her housework and cooking to do, as well as helping me. Any treatment of cuts, medicines or injections are done before she goes. The feeding of the babies is similar at every meal until all are weaned on to fish at about three weeks old. Then the feed is all whole fish, and the skinning, boning and emulsifying stops, so the work is a little easier.

In the summer the day's work alters. The preparation of the fish for the seals in the pool is still the same, but one of my staff sees to the filters, as I have to organize everything for the crowds of people visiting the Sanctuary. Talks have to be given to school and coach parties, also explanations to individual people interested in our work.

In the winter, as it gets dark by 5.00 p.m., we get all the feeding done, see to the filters, then close up by this time. Usually we have to come back around 10.00 p.m. to give the babies their last feed. But in the summer month the hours are continuous from 8 a.m. to 10 p.m., seven days a week, for Mary and myself.

CHAPTER EIGHTEEN

Rescue work

FROM September to March every year, when the weather is usually at its worst, we are on call 24 hours a day. The calls come from the Coastguards, Police, RSPCA and the general public and cover an area bounded by St Ives, Hoyle, Godrevy, Portreath, Porthawan, St Agnes, Perranporth and Newquay in the north and Mullion, The Lizard and Newlyn in the south. Occasionally baby seals are washed up in other little bays along the Cornish coast, often in spots which are almost inaccessible. Sometimes we have to drive across fields down steep slopes where the Land-Rover nearly turns over, or along tracks covered in thick mud where the wheels get stuck. Then there is the climb down slippery rocks with drops below of hundreds of feet. People ringing up about a stranded seal often give us misleading directions and we have landed up in some very queer places and had some funny experiences.

In October 1976 I was in the Hospital getting ready to feed three baby seals when the phone rang. It was a Mrs Bownas who owned a large guest house overlooking the beach at Mawgon Porth, and she sounded very worried. A baby seal had been washed in, and not knowing what to do she telephoned the RSPCA at Bodmin, but unfortunately the Inspector was out, so she rang the Police and got no help, so finally she rang the local vet, who told her to ring me, but also offered his services if the seal needed immediate attention.

When I told her I would collect the seal as soon as I had fed the babies in the Hospital she was relieved, but as the journey from Gweek to Mawgon Porth would take over an hour, I asked her if it was possible to get the seal off the beach to safety as the tide was coming in, and the baby would be washed out to sea again.

We had had many experiences of this happening, and in fact only the day before, a man on his way to work at Portreath saw an injured baby seal near the harbour. On receiving his telephone call we set off immediately, but on arrival the baby seal had vanished. During the hour's drive the tide had reached the spot where the baby was resting. A thorough search of the beach and rocks produced nothing. If only the people reporting stranded seals could get them to safety first it would prevent a wasted journey to save their lives.

Luckily, down at Mawgon Porth Mrs Bownas had managed to get the help of the local dustman, who took a dustbin down to the beach and carried the seal back in it to one of her bathrooms. It was put in an empty bath, and was safe until we arrived.

Paul, my son-in-law, came with me, with myself driving the Land-Rover as I knew the way. Mawgon Porth is about 5 miles outside Newquay on the north coast and the sea was quite rough there, but nothing to what it had been a couple of days before. We were surprised that we hadn't received many more calls. We found the guest house called White Lodge easily enough. Mrs Bownas and her daughter greeted us, offered us cups of coffee, and while we were drinking she gave us all the information. She then took us to the bathroom and there was the seal lying on a blanket in the bath. Mrs Bownas was really thrilled about saving the baby, and took quite a few photographs of it in the bath as a memento. She named it Samantha.

The journey back from Mawgon Porth was a rare experience. Having placed the seal into the back of the Land-Rover, and secured the hood so that it couldn't climb out, I let Paul drive on the return journey, so that I could keep an eye on the baby seal. About a mile outside Mawgon Porth, the effects of the drinks we had had worked on me, and a quick stop was necessary, so I told Paul to pull in at the first convenient spot. We were driving along by the aerodrome and pulled up near an old building suitable for my emergency. I jumped out quickly, and walked about twenty yards, Paul staying in the Land-Rover which was parked on a crossroads, just off the main coast road. I had just started to relieve myself when a Land-Rover filled with Air Force military police came tearing down the road, stopped at my Land-Rover, and two MPs rushed out. I was flabbergasted, the thought went through my mind as how anyone knew so quickly of my emergency, but I couldn't stop once I had started. Within seconds two police patrol cars came tearing up the coast road from the opposite direction, and out jumped four policemen. I was still not quite finished, when to top it all off, an ambulance arrived. I couldn't pull my zip up quickly enough, finished or not, and nearly needed the ambulance. It appeared that one couldn't even answer a call of nature without someone finding out, but for two MPs, four policemen, and an ambulance to intervene, seemed a bit too much. I nearly crippled myself in the rush and got back to find the Land-Rover surrounded by police. On conversing with the MPs, it seemed that a bad accident had been reported somewhere near the spot we were parked. After a while we continued our journey back, shaking with laughter at the thought of two MPs, four police and an ambulance converging on a man responding to a call of nature.

We arrived back at the Sanctuary safely and placed the seal into one of the pens in the Hospital alongside the three other baby seals newly rescued. The seal had no serious injuries, although it was very thin, but to our amazement it took half a fish on its own. I gave it more, but it had a job to keep them down, naturally, because it had had no food in its stomach for a long time, so I gave it a few half-fish every hour, and this it managed quite well. The following morning Mrs Bownas rang to see how Samantha was getting on, and I gave her the good news.

Paul, who had always been such a great help to me, would soon be leaving Cornwall with Linda to live in Birmingham. I was going to miss them both and would also need some assistance as the work was increasing.

So I took on a young man called Terry. He was short with jet black hair, and seemed a willing lad, so I thought I'd give him a trial to see if he liked it. First he worked on the filters and helped clean the pools, but his lack of height brought a few problems. When I asked him to get some fish out of the large chest freezer, it was nearly empty and while he was stretching to get the few fish at the bottom he fell in! In the filter house he had to stand on stones to reach inside. But he soon settled down, Paul showed him all that had to be done, and then he took over. The first baby was brought in on 22 September. It was 8 hours old, and the second, which came on 26 September, was about two days old. Both took to the bottle quite fast, so four times a day I had to skin, bone, and emulsify fish for the two of them. One was very small, prematurely born, and with inflammation over its tummy, which was very sore. Terry had never seen baby seals before so every chance he had he came up to the Hospital to look at them. He was fascinated by them, so sometimes I let him clean their pens out which enabled him to get closer to them. Other babies came in from the usual breeding places. In all we had 24 babies in the winter of 1976-7.

One had smashed its back flipper, had injuries

around its neck and body and was very thin and weak. We kept it in the front of the Land-Rover on my lap while Terry drove, as we didn't want to give it any more jolting than necessary owing to its bad condition. The flipper was swollen up enormously and must have been very painful. On arrival the pup was injected against infection, and also got a vitamin injection to help it along. When it came to force feeding it with fish I had to be very careful because of the flipper. This seal must have been about five weeks old and had lost its white coat, possibly it could have been washed away from its mother when two weeks old, and had since lived on its fat. As I held it in my arms his poor little head just dropped down, he was too weak to hold it up. I picked a very small mackerel, as anything large would have choked him. I'll never forget a seal in similar condition many years ago who brought the fish back and did choke, so I had to make sure that it didn't happen again. I gave him two small mackerel to start, four times a day, to get his stomach used to solids, then after three or four days increased it to about 5 or 6 mackerel. It was very difficult getting his head in a good position to push the mackerel down, as I had to keep any weight off the injured flipper. He made no resistance when I opened his mouth, I suppose he just didn't have the energy. I gently pushed the small mackerel down, and had to push it all the way as he couldn't help in any way and the fish slithered down on its own. He seemed to go breathless, but after a short interval opened his eyes and breathed again to my relief. I waited for the fish to settle and pushed down the second fish. I gently lowered him down on to the woodwool which I had put over the wooden slats to keep him warm. Also I put the infra-red lamp near him to give him a little heat. He just flopped over having no energy whatsoever. I cleared the woodwool away from his nostrils to make breathing easier, placed his injured flipper in a good position to make it comfortable – it was at least three times the normal size. The first six feeds were similar, he was very weak, we had to be very careful with him, but as the nourishment began to be absorbed properly he slowly raised his head when I came with the fish. He began to help a little with the swallowing of the fish, still giving no resistance to the force feeding. We injected antibiotics into the flipper, and later made a small opening to squeeze out all the pus. He used his other flipper to support the injured one, but when he started to move round the pen the injured flipper, which was still swollen, just dangled behind. I had seen seals with damaged flippers before but never as bad as this one; he could make no movement with it at all. Other seals could stretch their flippers or open them like a fan, his just lay limp. After three weeks the swelling went down, but his flipper never got back to normal size. He still couldn't move it, and can't to this day.

We removed the infra-red lamp to harden him off gradually. We were still force feeding him, but he was much stronger now. In fact, when I put him down after feeding he used to take a little bite at me. I didn't mind this, as when they start to bite you know they are getting better. Of the twenty-four pups which came in this winter, most were babies with a few weaners who hadn't taught themselves to eat. These were fairly easy to wean and we soon had them down in the pools. The two who had to stay the longest in the Hospital were the prematurely-born baby with inflammation on its tummy, and the pup with the smashed flipper. It was a busy winter fetching seals from all over Cornwall.

We didn't have many visitors through the winter, but those that did come enjoyed the bottle feeding, but were flabbergasted at the way the whole fish had to be pushed down the seals' throats, also at the amount of fish the young ones took, anything from eight to sixteen mackerel.

It was March before we took the seal with the injured flipper and its female companion down to the little pool, and as soon as they were both in the water, she chased him around the pool. He wasn't too happy in the water, and used to climb up the rocks on to the top of the small cave. She learned fairly quickly to pick fish up in the water, but he had to be hand fed, and still has to. He places his tongue in queer positions, instead of straight out to suck the fish in, he pushes it out to the side, so I have to hold the fish until it disappears completely.

At Easter when a crowd of people were around the pool where I was feeding him, they saw how he waited at the gate for me. In fact all the seals know their feed time. When I bring the Land-Rover down the hill, the seals in each pool jump out, or sit up in the water following it all the way down. The little one with a bad flipper comes up to the gate crying for his fish. I have a job opening the gate, and then as I walk to the other side of the pool, he'll flop along behind me crying as he goes along. He usually shakes his head first, then lifts it for the fish. He'll sniff it all over, then open his mouth and down it goes. I usually pick the smallest mackerel for these two invalid ones. While I am feeding the seal with the injured flipper with one hand, with the other hand I'll throw an odd fish to the female. She is very funny; she'll play with the first few fish, but as soon as I start to move to the next pool, she'll come swishing down the pool through the water. So I'll chuck one up the other end of the pool and she'll fly up after it, swallowing it straight down, and be back for another; in two minutes all the fish have gone. If I throw an odd fish on to the slabs she'll go slithering along after, pick it up, take it into the water and away again.

It was at this time the seal we had named Terry went off his food. He started to get thin, and even began drinking the pool water. Then one day I noticed his motion was black. I couldn't give him antibiotics in the fish as he wasn't eating, so the only way was to get him up to the Hospital for a few days until we could get him eating again. He had also turned nasty towards the other seals, so we emptied the pools, took the cage in, and transported him by Land-Rover up to the Hospital. We immediately gave him an injection for his liver and a suspected bleeding ulcer in the intestines, also a vitamin injection to try and help his appetite.

After his first injection he turned nasty on me. Previously we had been great friends, he always jumped out of the pool to greet me when I went in, and always accepted a good fussing. He had four injections in all, and after each injection I tried tempting him with fish, but he just snapped at me. I asked my assistant Terry to clean his pen out, but when he went in, Terry, the seal, chased his namesake round the pen, and I have never seen the human Terry move so quickly as he jumped over the rail. I cleaned it out myself after that as I didn't want anyone bitten. Normally he had such a gentle nature. On the fourth day in the Hospital, he responded better to the feeding, and after about ten minutes' persuasion he took the first fish. That feed he took about five fish, but immediately after he was aggressive again. Four times a day I continued this until he was taking about fifteen fish, I then cut him down to three feeds, but he was still very upset and poorly. He was in the Hospital twelve days altogether.

When we brought him up from the pool he hardly had a hair on his body, it was the moult period, but I had never seen a seal so bald. Most of the other seals had finished moulting and had nice new fur coats. However, by the time he left the Hospital, with his condition improved, his fur had practically all come back. He still wasn't himself, but at least he was eating well, in fact I couldn't give him enough. He had stopped his snapping at me after feeding, so I hoped his reaction with the other seals would be the same. When we got him in the water, he just swam up and down the pool ignoring all his mates.

I am often asked by visitors to the Sanctuary what I think of the seal hunting in Canada, and here in Britain too, in Scotland, the Wash and Norfolk. Approximately 180,000 seal pups die each year in Canada and their skins are made into fur coats for the rich and fashionable. Seal skins also make gloves, hats, key rings, and many other souvenirs that visitors buy when on holiday. A lot of people have told me of the beautiful ornaments made of seal skin that can be purchased in Norway.

In earlier chapters you have read of the suffering that baby seals have to endure: injuries, starvation, seagull attacks whilst they lie helpless on the beaches, rough seas smashing them against the rocks, pollution, getting trapped in fishermen's nets (accidentally) and drowning. In addition, because there are some women who refuse to wear the most realistic-looking fur fabrics, hundreds of thousands of baby seals are doomed to die a cruel death each year. What a pity one can't run for the owners of these clothes and souvenirs a film of baby seals being skinned alive, the screaming of the babies, the blood spurting everywhere, their beautiful eyes once full of life dimming in agony. If only they could be made to realize that their purchases help to keep this terrible slaughter going on year after year.

I have spent the last twenty years of my life caring for seals, being with them for 15 to 18 hours a day, seven days a week. I know their thoughts, their feelings, their sufferings. I will never forget a lady visiting the Sanctuary when it was at St Agnes. A baby seal a few days old had just been brought in, with beautiful big eyes, and its unblemished white coat glistening. Chatting to the lady she told me she had a seal skin coat, and a pair of gloves, but wanted a hat to match. She had tried the Ministry and various other places for a seal skin to make the hat, but had no success. I was so shaken that I took her into the pen to see the baby, thinking it might help to change her mind. When she saw it her eyes glistened and she turned to me and said, 'Oh wouldn't it make a beautiful hat!' I was speechless, the baby seal toddled over to us giving a little cry as a normal baby would, I stroked it, and in my mind I imagined this beautiful little creature being skinned to make the woman in front of me a hat. I hurried the woman out quickly with tears in my eyes. My tears were not only for the lovely baby seal but for that woman who even after seeing the helpless baby there in the pen still wanted a hat from its skin. I realized that no amount of talking would persuade her differently, so I had no further conversation with her. I walked away and went into the house to recover over a cup of tea and a cigarette.

Seals have been around for thousands of years, and no mass killings took place until recently. This century, when man's greed is becoming worse each day, hundreds of thousands of seals have to die and I suppose when the seals are extinct, they will find other animals to slaughter, until finally only we are left. Most seals live in sea water, but they breed on land, suckle their young, and then mate again. It is during the breeding season, while thousands of mothers are suckling their young, that they are most vulnerable. It is very easy then for the hunters to rush in amongst the vast herds, frightening the mothers into the sea, and leaving the babies to the peril that follows. It seems that the voices of millions of people protesting these cullings count for nothing, against a few hundred people who make a living out of it. The only way is for people throughout the world to destroy the market for sealskin products, which will in turn end the slaughter.

Over 50 per cent of seal pups die each year before reaching the age of six weeks. These are natural deaths. In addition there are natural predators, polar bears, whales and certain sharks who eat baby seals, and this accounts for further losses. And taking into consideration the massive factory ships now fishing the seas to extinction, seals in the future will be facing starvation. The whole balance of nature is being upset, and in years to come we will pay for it, unless the people of all countries band together and stop the destruction which now goes on. Of course fur dealers, fur factories and certain fishermen make the excuse that it is done for the benefit of the seals, to keep populations down. What we do know is that they are the ones making money out of it. Many seals of various types have already suffered the fate of extinction, so let us see that no further species suffer the same fate.

Seals live a hard life, especially the first six or eight weeks of their lives. From the day they are born their lives are at risk, with the seas pounding their breeding grounds, smashing the babies against the rocks, or washing them out to sea where they get exhausted and drown. They sustain terrible injuries, and many are lucky to survive at all. The mothers do everything they can for the babies. Even when the hunters come the mothers try to protect their babies, and very often in protecting their young suffer the same agony the babies do, by being bashed on the head with a club.

In my Hospital I have a large photograph of a Harp baby seal just about to be slaughtered. In the pens lining the same wall are ten baby grey seals of the same age, just a few days old. The one on the wall is now dead, and possibly being worn on the back of a fashionable woman, the ones in the pens are alive and will be ready for a life at sea.

Large groups of people gather together each year to fight these terrible slaughters in different parts of the world, but more support and financial help with the costs involved is needed. It is not enough to sit and watch the killing on television and feel sorry, that doesn't help the seals, but joining the various organization and actively campaigning does.

Excuses are made that if a cow or bull can be killed, then why not a seal? A lot of people are vegetarian, and they would question that morality. Even those who do eat meat will say that these animals are bred solely for food, that food is essential to life and that fur coats and ornaments are not.

I have risked my life climbing down cliffs and steep slopes to get at injured baby seals, been bitten whilst force feeding them in the Hospital, but the joy and satisfaction of seeing those babies, strong and healthy, returning to the sea, ready to cope with the hazards in front of them, is a great reward, as is the knowledge that although more will suffer the same injuries each winter, help is now there for them when required.

Unfortunately in the world today people are getting selfish and greedy. Until we help each other, and all the creatures God put on earth, we shall never achieve the enjoyment and fulfilment we were put here to obtain. To watch families of seals playing in the water, and on the sandbanks or rocks, is like watching a young couple with their children playing on the grass, all enjoying life, but to watch a seal clubbed and skinned whilst screaming in agony is like watching children in a bombed building, this all brought about by the greed of man. Unfortunately, baby seals are so trusting, even to man, and their big eyes shine up at you with no knowledge of the cruelty within the human race.

It is hoped that in the near future, sufficient pressure will be put on all Governments all over the world, by the mass populations, to stop this mass slaughter of baby seals. I hope in the coming years that we human beings will change a lot of our greedy practices: first by refusing to wear clothes made from the furs and skins of animals, then in the fishing industry to remember that other creatures live in the sea, who are wholly dependent on its contents for life. I hope they will not continue to scoop everything out of the sea for a short period of high profit, leaving the future barren for us and destroying the food of other sea creatures. I feel that the way fish is being caught at the moment and used not only for food, but for fertilizers as well, the future of all life at sea is in peril. We must learn to preserve what nature has given us, and think of others for once instead of ourselves. Wildlife is as much part of the universe as we are, and I often think they would take better care than we do.

If you see things like cruelty then do something about it, help fight any slaughtering of wildlife done just for profit, don't just sit back and feel sorry for the animals killed. The more pressure that is put on, the quicker it will stop. We were all put on this earth to enjoy what is on it, and to enjoy each other's company. Instead all over the world there is killing of humans and animals — either for profit or sport. One cannot change the world in a day, but we can all make it a better place to live in. There is much more pleasure in watching the flight and action of all types of birds than shooting them.

In my life I have tried to give what help I could to all sea creatures, and sometimes those inland. I hope in the future more people will do the same and that it will be encouraged by future Governments as a lead to the world to protect what little we have.

CHAPTER NINETEEN

Our first baby seal

SEALS do not start mating until the female is at least six or seven years old, and as some of the residents in the large pool had been with us for a number of years they were now coming up to, or had passed, mating age. I had had a serious talk with all my resident seals the previous winter, explaining the facts of life and telling them that I would be getting a lot of babies washed in so they must be careful not to give me any more. They'd all promised to heed my words but now both Jenny and Sheba were in an advanced state of pregnancy! I pretended to be cross that they had disregarded my warnings but of course we all looked forward to the happy events with great excitement.

Seals walk on their bellies so that their sexual organs are not at all prominent but naturally quite a few visitors are interested in their mating habits, especially if they had happened to witness a bit of love-play between males and females on the slabs. Even the pups too young to mate like to cuddle up together and it is very pretty to watch them as they use their flippers like little arms. I remember a young policeman and his wife who were very interested in the details of mating. When I had finished explaining the wife was so interested she said she would like a job at the Sanctuary. The next day she brought her seventy-year-old father along and I had to go through the whole explanation again with him – he was even more interested than she was!

The female grey Atlantic seal breeds in November after an eight- or nine-months gestation period, very like the human one. After feeding the baby for three or four weeks after birth, the female is ready to mate again. But this is the only time in the year that she is. They mate and breed every year until they die and as the life of a seal is around forty years they can give birth to about thirty-four babies in their lifetime. There is no age limit for motherhood and recently the Seal Research Unit recorded a forty-year-old seal being washed in who was expecting a baby. You can see that I'm likely to be in for overcrowding problems as the years go by and I might have to institute a bit of family planning by separating the older seals into all-male and all-female pools.

By September Jenny and Sheba were huge and I calculated that they would have their babies in October or November. Fortunately it was a slack time for visitors so we were able to apply ourselves to the task of moving the pregnant mums to one of the small maternity pools. We wondered how on earth we were going to carry them even the hundred yards from the large pool to the small pool. Even more, how were we going to lift them out of the large pool, as they weighed over 700 lbs each?

We emptied the pool, got out the big cage, then had to try and gently get them in without upsetting them too much, as they were so close to giving birth. Terry and a friend of mine, Ken Ashford, helped me. We managed to drive Sheba and Jenny one after the other slowly into the cage and pull down the sliding door. Even three of us together couldn't lift the cage, so we called on some men looking on to help. 700 lbs seemed like seven tons, but eventually we got them out after a tremendous struggle, all of us puffing and blowing like mad. Then we edged the cage three feet at a time along the top of the pool to the gate. By the time we got it down to the small pool we were completely exhausted, and decided then that something different would have to be done to get them back at a later stage. Jenny and Sheba weren't too happy in the little pool, but it would only be for a short period. We knew they were close to their time but couldn't tell the exact day, so we were watching for any signs. They didn't eat a lot for the next few days, but they could hardly move as it was, so I didn't worry too much. On 3 November Sheba was sick, a sign that it would be any time now. Suddenly a piece of the afterbirth showed, and I knew this meant trouble. She was now beginning to press, so anything could happen any time. One of the few visitors at the Sanctuary was a doctor, and he confirmed that we were in for trouble. We decided if necessary we would stay with her all night. By nine p.m. the bag was showing about six inches and it wouldn't be long now; she was pressing hard, but finding it difficult. Sheba had very bad lungs, her breathing was very heavy, and at intervals she had to have a long rest. She was still in labour at four o'clock next morning; I was tired and thirsty, so I thought I'd go for a short rest and a cup of tea. At 6.30 a.m. I came back and saw that the bag was a little farther out, but things didn't look right. I was just going in through the gate when out the baby came. It was dead. Poor Sheba, she had had a dry birth, and this had been the reason for all the straining. It must have been what the doctor meant when he said we were in for trouble. I felt really sad to think she had carried it for so many months and just when she was to become a proud mum, something like that had to happen. I spoke to her softly as she licked it, comforting her in the way any mother would want to be comforted at such a sad time. Sheba and I looked at each other, but I couldn't tell what was going through her mind. I nearly cried; I know these things happen, and can't always be avoided, but it's sad to see a perfectly formed and beautiful baby which a few days before must have been ready for what the world had to offer, lying there lifeless. However, I had to take some quick action so I fetched a plastic sheet, wrapped it round the dead baby and took it away. Ken Ashford had proved a great help, throughout. He came down, with Terry, in his spare time (he worked at Culdrose Aerodrome) to keep watch, helped in all the carrying and felt as involved as I did at the loss of the baby. Poor Terry was so upset that he couldn't watch any more, as Sheba gave birth to her dead baby. I think it made him sick, and I couldn't blame him, it was a sad event.

I came back to Sheba straight away to console her and to make sure she was all right. She was sniffing around for her baby as one would expect, but looked completely exhausted, after the very difficult birth. I hoped that it hadn't taken too much out of her and I kept checking on her condition all that day until late evening. When at last I saw she had settled down again, and her breathing was better, I turned in for a much-needed sleep.

It made me wonder if the same thing might happen to Jenny. Three days later she began to show, and in only two hours out popped a beautiful healthy baby. She sniffed it, licked it, and then pawed it as if she was pulling it towards her. The pup nuzzled into its mum's tummy, searching for the teats. A seal has two teats low down on the tummy below the belly button.

It took ages for the baby to find them, and Jenny just lay there patiently, still using her flipper to encourage it. Then suddenly it found a teat and started sucking away merrily. I cleaned up all the mess, and the afterbirth, and all was well. The pup doubled its weight in the first week, and at two weeks must have weighed one hundred pounds. At this stage I removed the pup to the Hospital, and put Jenny and Sheba back into the residents' pool. But this time we made a special wheeled trolley to put the cage on, to save the effort of carrying it, and it only took a few minutes to get them back to the pool. The baby weaned quickly on to fish and was much bigger than most seals of the same age. This was going to be a healthy pup who would be in good strong condition when he was put to sea. I thought how lucky he was compared with the others who faced the pounding waves, cruel rocks and all the other perils of the sea from the moment of birth. But we were going to be very sorry to part with our very special baby, the first to be born in the Sanctuary.

CHAPTER TWENTY

Donkeys, ponies and horses

A FEW months later, in early May, we were expecting another 'happy event'. Meemy, the donkey who had been brought to me for grazing, was due to foal. We put her in a small paddock near the pools so that we could keep an eye on her. She had a queer nature, butting you with her head, then with the sides of her body; she'd rub against me and knock me over. She did this every time I took her some sugar and potato peelings and carrots from the house. The other donkeys, Manny and Minny, were in a separate field as we didn't want any problems with the baby donkey when it was born, as they were very jealous.

On the Thursday of the first week in May I checked on Meemy to make sure she was okay. That was at two o'clock. I had another peep at four o'clock and there it was – a dark brown baby with very thick fur. It was a big baby too, a female and having all it could do to keep on its feet. Cameras clicked as visitors came down to see it, the mother was very proud and keeping the baby close to her. It had a job finding the teats and took a good hour before finally getting its head right under. Meemy did all she could to help it. The watching visitors were exclaiming how big she was, and how cuddly she looked. It was a healthy baby, and Meemy was in perfect health too.

The next few days everybody had to look at the baby and each day she grew stronger on her legs, and followed her mother everywhere. She loved having her neck stroked, in fact she got very cheeky and always pushed herself in front of her mother to have some fuss.

In no time at all she was chewing at the grass, then she got frisky and used to kick up her back legs at her mother. She was really being spoilt. She seemed to grow quite fast, and was also getting much quicker in tearing around the field. The paddock was very small, and the grass had begun to get pretty low, so I thought I'd put them in the bigger field with Minny and Manny. I wasn't sure what their reaction would be, as these two had always been on their own, and were inclined to be jealous. The trouble was that Meemy was a devil to move, so we used the baby as a decoy. We had about two hundred and fifty yards to take them, and a choice of two ways; either up the tarmac road, or through another field, then across the road into the field with Minny and Manny. We tried leading Meemy, but she stood as firm as a rock, moved a few yards, then stopped again. As the baby was friendly with me, I had her follow me through the opening in the fence, and naturally Mum came after.

Then the baby got frisky and bolted right up the field with me chasing her. I thought if I could chase her in the right direction we might get somewhere, but, no, she had to go in all directions. I was running around like a fool, and she made me look like one. Away she went up the field, with Meemy slowly trudging along the fencing, having to stop occasionally to chew some grass. We got them through the gate, which left about 30 yards to go; now we had to cross the tarmac road, then straight into the field. I just couldn't get the little one across the road. Finally I had to practically carry her and push her straight in, then we closed the gate and waited to see what would happen with Minny and Manny. They came charging up the field to the baby, sniffing her; she kicked out her back legs, and bolted around the field, Manny and Minny chasing her. Mum trudged slowly behind. I went to Minny and Manny and gave both of them a good fussing, then I did the same to the baby, and back to them again. After this they seemed to settle down, but occasionally the little one would frolic and back kick.

The only time I had trouble later was when I took sugar and peelings. I had to put them in two piles, one for Minny and Manny, and one for Meemy, the baby had only sugar. Even then they would tend to go for each other's piles and Meemy would soon turn her back and kick.

In the pools inside the Sanctuary we have our family of seals and in the fields just outside our family of donkeys and ponies. The ponies are Toh-Toh and her daughter Sunshine who is growing so fast that she should make sixteen hands, much bigger than her mother. Then there is Little Blue who looks like a little lamb in comparison, following in their tracks, and Pippa, a young grey stallion who chases them all around the field and sometimes chases us as well and nips our bottoms as we walk across the field. Minny and Manny, now joined by Meemy and her new baby, have to have three lumps of sugar and they all look forward to their daily ration. The donkeys also have potato and carrot peelings when available and they all have a few minutes fussing from me every day and a quick brush down. They are a great attraction to the visitors who gather round with their cameras at the ready.

Everyone admired Sunshine, and a few people have made offers to buy her, but I was not interested as I intended to break her in to harness to drive one of my horse-drawn vehicles. This is one of my latest ideas and one I hope that will give extra pleasure to visitors. About ten years ago I was given an old cart in very bad condition. In my spare time in the winter evenings I worked in the garage renovating it, replacing rotten wood, stripping off all the paint and repainting with six coats of paint, each coat of paint being rubbed down before putting on the next. As this was my first attempt I naturally made many mistakes, and although to me it looked good, it didn't pass the scrutiny of some of the members of the Vintage Horse-Drawn Vehicle Club. It took me another three years to learn thoroughly the art of restoring these vehicles to their original high-class condition. There is a terrific amount of hard work in renovating even a two-wheeled vehicle, and naturally a four-wheeled one takes twice as long. There is all the woodwork to be renovated, the uphostery repaired, all the metal work, axles, springs, supporting iron work, turntable, wheels and shafts put in working order. Some have to have quite a lot of leather work replaced, although if possible the old leather is retained.

Horse-drawn vehicles are graded in stages in the way cars range from Minis to Rolls-Royces; the cheapest is a jingle, then a raleigh, gig, phaeton and dog cart, with the grandest a landau like the Queen uses. But the owner of such a vehicle needs a horse, and with the horse goes a harness for pulling the vehicle, so they can be more expensive than a car to buy, but much cheaper to run. Also, they're much more fun. Over the years friends have given me old vehicles which have been abandoned on farms and left to rot away. Some, I think, were given because they had seen the work I was doing on those in the garage.

All horse-drawn vehicles were painted in special colours, and not in the bright colours of gypsy caravans. The lining is delicately fine and in the old days it was done with a long-haired fine brush. It is said that the more drinks the man who did the lining had had, the straighter and finer the line, his hand got steadier. When I first tried it you never saw so many waves in a line in your life, but patience and experience soon changed that. The same with the painting, anyone can hold a brush and slap paint on, but to get a finish as smooth as glass, a lot of work has to be put in. The wood must be rubbed down, all holes filled in, then each coat of paint when dry rubbed down with wet and dry paper until really smooth. It is this build-up that enables the final coat of paint to look as if it has been sprayed on. Then the contrasting colour for lining puts the finishing touches, and gives the smart, gleaming look that everyone admires.

No one looks twice at a car passing on the road, but a horse-drawn vehicle always attracts attention. The club I joined used to have drive-outs every month. About ten or twelve of these horse-drawn vehicles driving along the main roads brought crowds of people, and any money collected was given to charity. All the women passengers dressed up in old Cornish costumes and looked very smart. With vehicles cleaned up, harness polished, and horses groomed, the turn-out looked really splendid.

In the summer the horses and vehicles attended shows all over Cornwall: gymkhanas, horse-jumping shows, the Royal Cornwall Show, and local carnivals. In some instances they were used for weddings.

I didn't have time to go to these shows myself, but my vehicles were used by friends. Now I intend to use them on my own land, giving people the chance to ride in these lovely old vehicles (some date back to 1890) to and from the Sanctuary. I have placed two hay wagons already down at the Sanctuary, but the better vehicles must be stored under cover if they are to be kept in good condition. How nice it must have been in the old days with just the clip-clop of the horses' hooves instead of noisy motor-engines and petrol fumes! One of my vehicles is a governess cart that was used by a doctor at the turn of the century in Cornwall. How interesting it would be to go over some of the journeys he made. Each year the vintage value of these vehicles increases, and it seems a shame that so many of them were left to rot in barns when the motor car took over. It's nice to keep a little of the past, when so many of our English ways are disappearing.

However, before the carriages could be used, horses and ponies had to be broken in to harness. We started by sending Sunshine for training and decided to use the opportunity while she was away to have Pippa, the grey stallion, gelded – we booked a date with the vet. Sunshine and Pippa had always been special companions so you can imagine the problems we had when they were parted. Getting Sunshine into the horse box and at the same time trying to keep Pippa and Little Blue from following was a real job. Even when we did get Sunshine into the box, Pippa got over the fence and galloped down the road chasing it. It took four of us to get him back, and we boarded up the hole in the fence. He was frantic and remained like this all day, so we had to keep a close watch on him. The next couple of days he just hung about the gate, crying out for her.

The vet came early one morning to do the operation on Pippa. We hoped it would quieten him down a bit, and he would leave the mares alone. This was the first time I had ever seen a stallion gelded. I held Pippa while the vet injected into a vein, and in seconds the pony was knocked out. The operation itself took about fifteen minutes, and Pippa felt nothing. The vet had a good sense of humour, and you can imagine the type of jokes that were flying about. He told us, when he was injecting Pippa, that if his hand slipped, and the needle went into one of us by mistake, that it would knock us out instantly and possibly kill us. As he was finishing the operation Terry came up, and the vet asked him if he wanted a transplant.

After the operation another injection was given, and within seconds Pippa was up on his feet, a little groggy at first, but soon on his way. I thought it was marvellous, that such a large animal could be knocked out so quickly, and back on its feet just as quickly. For the next few days we kept watch on him in case of bleeding, but all was well.

He did look sadly at me, but it wasn't only the trauma of the operation, but the fact that he'd lost Sunshine as well. Little Blue stayed close to him, but Pippa still hung around the gate looking for Sunshine. After three weeks she came back broken in for riding. She gave a lovely ride and really quiet, but had lost a lot of weight, through fretting as well as going through the breaking-in period.

Pippa was taken to another field as Sunshine was brought in, as he was going into the same horse box she had just left to be broken in the same way. He had healed up nicely, but they were going to take it easy with him just the same.

Sunshine now had to be ridden each day to improve her, because we had decided to enter her in a show at Constantine Gymkhana a few miles away. This was to get her used to other horses, and also to get a professional opinion about her. She looked forward to her rides, except when I got on. I had never ridden a horse before in my life, so this was going to be some experience.

I had my first lesson at seven o'clock one evening, when there were no visitors in the Sanctuary, and all the work had been done. Friends of mine had been riding her regularly, so they saddled her up and all was ready. It was like climbing a mountain to get on her back, I put my feet in the stirrups, got hold of the reins, and was all ready to go. Nothing happened, she just stood there with me shouting 'Get up there'. It transpired that I was holding the reins too tight, so I loosened them off and wondered what I was going to hold on to now they were slack. My mate touched Sunshine's bottom and spoke to her and off she went in a way that seemed to me like speeding, but in fact she was only walking slowly. It was like being on the top of Mount Everest, but I was still on her back. She kept having a glance back at me as if to say 'What the devil are you doing there?' My mate shouted 'Canter her', and showed me how to do this by touching her with my feet. She started to move faster, and I was bobbing up and down like a ball on an elastic string. I felt like Cowboy Charlie crossing the Prairie, except that I was frightened to death, I kept shouting 'Whoa, girl', but she kept on, and I thought at this rate the black bottom I had received from Spitfire would turn into a red bottom. I seemed to have travelled miles, but in fact I hadn't done a hundred yards. I managed to turn her round and start back towards my mate, who was still shouting 'Canter her'. I shouted back, 'Not ruddy likely! I'm going fast enough', and just plodded slowly back to them taking in breaths of fresh un-polluted air at this high altitude. He told me to take my feet out of the stirrups and when ready, to give a quick jump off. My twenty-first birthday is many years past and I looked at the drop to the ground, and thought 'Oh dear me'. But I did as he said, and actually made it. At least I had acquitted myself not too badly and Sunshine had been as good as gold throughout it all, so we knew we could trust her.

Little Blue was rather different. He had been ridden previously but for some reason had been treated badly. He was difficult to catch, but very good when he was caught. We decided to try the saddle on him, letting his good eye see everything that was going on, so as not to be frightened by sudden noise. There was very little trouble getting the saddle on, or putting the bridle and bit on. Terry, who was the smallest and lightest, got on his back. Little Blue was quiet at first, and then gave a couple of bucks. It was like watching at a rodeo, then suddenly Terry was thrown head over heels. From what I have learned about horse riding, one should get straight back on, or the horse will think it has won the day, and do the same next time. But what we did was to let my friend's daughter get on his back and then we led him round the field. This way he was perfect. But as soon as we put Terry back up, he was like Bronco Billy again, and off Terry came. The next few days we noticed Little Blue wasn't scared of children at all, but if we tried to catch him he was away. My friend Bob Ives had a son, who was very small (only about 3 ft) and very nice-natured. I call him Charlie. One evening he went up to Little Blue, held on to his halter, and walked him round the field. The horse followed him with no trouble at all. Charlie fell head over heels in front of Little Blue; the horse stopped dead, and put its head down to the boy. This, it seemed, was the answer: only children could really handle him, he only had trust in them.

It was nearly time for the first Horse Show and Sunshine was getting her final training and being groomed for the big day. She also had to get used to being placed in a horse box, and backing out of it. She wasn't too keen on this, but gradually got better at it. On the morning of the show, she was groomed up and ribbons put on, then put into the horse box. My friend Bob, with his wife and children, were taking along their own mare to the show.

CHAPTER TWENTY-ONE

Conclusion

THE number of visitors to the Sanctuary has trebled in two years – from 50,000 to nearly 150,000. Without Ken and Terry (plus extra helpers in the busiest summer months) I would not be able to cope.

Ken got so interested in our work that he recently gave up a good job at Culdrose Aerodrome to join the Sanctuary full-time. He comes from up-country but has lived in Cornwall for about four years – separated from his wife, he is bringing up their two children on his own, and making a good job of that too. Ken is a morris dancer and on his final day at the Aerodrome he was presented with a drawing by an ex-RAF officer which is one of the nicest and funniest I have ever seen. It shows Ken dancing in his Morris outfit with bells on his legs, seals lying in bed with plasters on them, the whole picture displaying him entertaining the sick and injured seals.

We all now work as a team. In the winter I shall teach them how to pick seals up, how to handle them for force feeding, and anything else necessary. Up to now my health has been good, but one never knows what can happen, so someone must be able to take over. Ken soon learned to cope with the pool system, the filtration, pumping in from the creek and the treatment of the water, and that is a tremendous help. Both he and Terry help to keep the place clean, are interested in everything that goes on and willing to turn their hands to any job that may crop up. They certainly look rather odd together as Ken is over six feet tall and well-built. As Terry stands not much over four feet you can understand that he's hardly likely to pick an argument with Ken very often. As well as the four permanent staff (myself, Mary, Ken and Terry) three very nice ladies look after the kiosk in the car park during the summer months, working in shifts.

At the moment of writing (July 1977) we have eighteen seals in the pools. The sun is scorching down, the water is in the creek, all the trees are blossoming, boats chugging up the river, birds chirping, people smiling, what more could one want out of life? All the seals are eating well, Flipper still talks for his food, Silky jumps out of the pool and pinches a fish out of the bucket, Blacky's breathing is still very bad, Nelson has fallen out with his girl-friend, so he stays mostly up the top end of the weaning pool, Spitfire still jumps out for his fish but only eats the tail end of the fish, and I still tease him. Lucky (the seal with the cleft palate and hare lip) is eating well and is very happy, he is still being adopted by the Benson School at Oxford. Terry has been off his food again, but is now eating well, though I am afraid that I might have further troubles with his health. The baby with the smashed flipper is eating well again, but still has to be hand fed. It still cries for its food as I come with the fish bucket. Jenny and Sheba are getting fatter every day. They are 'expecting' again but I think it will be early December this year before they have their pups. All the donkeys are doing fine, especially the baby. Sunshine came back from the Gymkhana without winning any prizes. She had lost too much weight during her breaking-in period by fretting, so we shall fatten her up now and not worry too much in future about shows. She is a pet and will be used just for riding around the area, which we shall both enjoy.

In June Mary and I celebrated our Silver Wedding – we had a little party and invited all the seals. It is the greatest joy for a man to have a wife prepared to help him in his life's work, support all his endeavours, and help him fulfil his dreams. I gave thanks for my good fortune, also for our wonderful daughter, Linda, and her husband Paul, who had never failed to come to my help in all sorts of situations. Without Mary, Linda and Paul I could never have achieved as much as I had.

My birthday was in June too and looking back over the years I realize that I have been one of the lucky ones of this world. Very few people ever have the chance to get close to nature and understand it but fate has given me the opportunity to care for animals, to understand their feelings and their difficulties, to give love and get back love from them.

For years I worked far underground in the mines and learned how hard, dangerous work and the dark pits could be lightened by the spirit of close comradeship. I also learned from the rock formations miles below the surface the changes that had taken place over millions of years – fossils of fish from the seas that once covered the land to be replaced by forests which in turn were converted into coal which enabled our country to prosper and grow rich.

Later as a hospital worker, I learned how to care for the sick and mentally handicapped, to understand all different types of people and to have patience and love for others. When I was running my beach café business I learned how to cope with frustration and worry, to endure long, tiring hours with little time to eat or sleep. Voluntary Civil Defence and Coastguard duties brought me a different type of knowledge and all these various experiences finally led me, better equipped both mentally and physically, to my destiny, the founding of the Seal Sanctuary. Looking back, I often wonder why I was chosen for this work but I'm thankful I was and only regret that the few seals we lost couldn't have been saved.

I shall carry on fetching and handling the seals until old age or poor health stops me. Then it will be the turn of my younger staff to take over but I shall still give advice if needed. We have now established a first-class organization for rescuing and treating injured seals and each year aim to improve the system. Thinking of my twenty years as a 'seal doctor' and the hundreds of seals that have passed through our hands the ones that I'll remember most are Simon and Sally, Judy, Kim, Silky, Lucky, Spitfire, Nelson, Terry, Benny, Flipper, Sheba and Jenny.

I like to think that our work is not only caring for seals but in educating children (and adults too) to understand a little more about our environment and the preservation of wild life. I hope that all the people who have visited us over the years have learned something valuable. I wish to thank them, for without such support, the Sanctuary cannot survive. If you who are reading this book haven't been to see us yet, then plan it for the future. We will be delighted to see you.

Afterword

A lot has happened since 1977. The planting around the Sanctuary has matured, the pools now look part of the landscape. Many thousands of people have visited the Sanctuary and many seals have passed through our care.

The death of Lucky, the seal with a cleft palate, who had been adopted by the Benson School in Coxford for most of his life, was a recent and great sadness to us. He was always difficult with his feeding, some days eating well, other days just playing with the fish, and in the week he died he hardly ate at all. He always had bad ear infections and it was my intention one Friday to take him out of the pool and to the Hospital for an injection.

I got up early one Friday morning and went to the resident pool which Lucky was in. At first I thought he was sleeping, so I shouted to him. Normally he lifts his head, but he didn't move. I jumped down into the pool and the other seals (Flipper, Benny, Spitfire, Dracula and Fatty) came towards me. Lucky still didn't move, so I gently touched his back and found he was cold and stiff.

I couldn't believe it, because we had been through such a lot together over the years. When we first had Lucky, Mary and I had to forcefeed him which was very difficult. It felt funny pushing a fish down his throat, because he had no nostrils as such, and on the first day we put him into the pool he couldn't stay under water like the other seals. He was approximately twelve years old and I suppose he was lucky to have survived at all: it had never been known before for a seal to have a cleft palate. We shall never forget him, and I'm sure that thousands of other people will remember him also.

Over the years we bred many baby seals, in fact so many that it became difficult to accommodate the pregnant mothers owing to a shortage of pools. In addition, many dolphins and killer whales had been washed up on the beaches. We couldn't care for the larger mammals – the Sanctuary was much too small – so they had to be shot.

The planning authorities were once again causing me problems, making things very difficult for me to do my work properly. In the end they put eight enforcement notices on me for various things, which meant the Sanctuary might have had to close. Greenpeace were called in, together with a top solicitor, to fight the planners. We went to a public appeal, which lasted four days. Thousands of people signed our petition against the planners and finally the Inspector for the Department of the Environment passed seven out of eight of our appeals. In all we were granted plans for four maternity pools, and a large pool with glass sides for viewing under water – this to accommodate the injured dolphins and young killer whales for a short stay while receiving treatment. In addition, permission was granted for a large lecture room which was badly needed for educational purposes; for schools, WIs, students and many other organizations.

Over the years I have also built up a large collection of Victorian items which is now on display in an old school at Mullion, near the Lizard, and money raised from this will go towards building the new pools and lecture room.

There are now thirty seals at the Sanctuary. September is close so more will be coming in and our hard work starting again. Come and see us again soon.

August 1985